Agricultural Extension

Second Edition

A.W. van den Ban
Formerly of Department of Extension Education
Agricultural University
Wageningen
The Netherlands

and

H.S. Hawkins
Faculty of Agriculture, Forestry and Horticulture
The University of Melbourne
Parkville, Victoria
Australia

Blackwell
Science

Editorial Offices:
Osney Mead, Oxford OX2 0EL
25 John Street, London WC1N 2BL
23 Ainslie Place, Edinburgh EH3 6AJ
350 Main Street, Malden
MA 02148 5018, USA
54 University Street, Carlton
Victoria 3053, Australia
10, rue Casimir Delavigne
75006 Paris, France

Other Editorial Offices:

Blackwell Wissenschafts-Verlag GmbH
Kurfürstendamm 57
10707 Berlin, Germany

Blackwell Science KK
MG Kodenmacho Building
7–10 Kodenmacho Nihombashi
Chuo-ku, Tokyo 104, Japan

First published in Dutch by
Boom-Pers 1974, 1985
Modified English edition co-published
by Longman Scientific & Technical 1988 and
John Wiley & Sons, Inc.
Second edition published by
Blackwell Science 1996
Reprinted 1997, 1998

Set in 10/12 pt Times
by DP Photosetting, Aylesbury, Bucks
Printed and bound in Great Britain by
MPG Books Ltd, Bodmin, Cornwall

DISTRIBUTORS

Marston Book Services Ltd
PO Box 269
Abingdon
Oxon OX14 4YN
(*Orders:* Tel: 01235 465500
Fax: 01235 465555)

USA
Blackwell Science, Inc.
Commerce Place
350 Main Street
Malden, MA 02148 5018
(*Orders:* Tel: 800 759 6102
781 388 8250
Fax: 781 388 8255)

Canada
Copp Clark Professional
200 Adelaide Street, West, 3rd Floor
Toronto, Ontario M5H 1W7
(*Orders:* Tel: 416 597-1616
800 815-9417
Fax: 416 597-1617)

Australia
Blackwell Science Pty Ltd
54 University Street
Carlton, Victoria 3053
(*Orders:* Tel: 03 9347-0300
Fax: 03 9347-5001)

A catalogue record for this title
is available from the British Library

ISBN 0–632–04053–X

Library of Congress
Cataloguing-in-Publication Data
Ban, A.W. van den.
[Inleiding tot de voorlichtingskunde. English]
Agricultural extension / A.W. van den Ban and
H.S. Hawkins – 2nd ed.
p. cm.
Includes bibliographical references and index.
ISBN 0–632–04053–X (alk. paper)
1. Agricultural extension work. I. Hawkins,
H.S. (H. Stuart), II. Title.
S544.B2513 1996 96–10929
630'.715–dc20 CIP

Contents

Preface

We are happy that readers have shown a lot of interest in our book. So far over 55 000 copies have been printed in eight languages. However, a textbook only remains valuable when it is modified to take account of new developments. Therefore this new edition had to be considerably different from the previous edition, although the basic ideas remain the same.

One of these basic ideas is that there is not one optimal approach to extension, but that the approach depends on the goals of the extension organization and the situation in which it operates. Hence, we do not give recipes which have worked well in one situation and might not work at all in a different situation. We try instead to present basic principles which can help readers to analyse their situation and to find an approach which works well in this situation.

There are several reasons why it was necessary to introduce changes in this new edition. Firstly, we have learned from our experiences, one as a consultant and the other as a teacher, how this book could be improved. We are grateful to the farmers, extension agents, students, extension scientists and others who have helped us in this learning process, and especially to our colleagues at our universities.

Secondly, agricultural extension is one of the policy instruments which supports agricultural development. Thinking about agricultural development has changed considerably as the role of government decreases in many societies and the roles of private enterprise and of non-governmental organizations increase. Economic liberalization also has created new opportunities for farmers, while at the same time exposing them to more competition. This has made it more difficult to predict the future and to advise farmers what is the best path for them to follow.

Finally, many new ideas about agricultural extension have been published since the first English edition of this book. This is partly because of increased recognition of the importance of environmental problems and sustainable agriculture. Also more recognition has been given to the value of indigenous knowledge and knowledge developed by farmers' own experimentation and of their participation in the development process. Furthermore, privatization of extension services has stimulated new developments.

It was not always feasible to integrate all of these ideas in our book, because they are sometimes contradictory. We also feel that some people who introduce a potentially valuable new idea exaggerate its importance. We do not try to convince our readers to accept these new ideas, but we wish to help them to make up their own minds about the value of these ideas for their particular situation. That is why we refer the reader to many of the original publications. We also have had to delete several parts of the previous edition in order not to enlarge the book too much.

Some readers have suggested that we include the methodologies of Farming Systems Research, Rapid Rural Appraisal, Participatory Rural Appraisal and similar research methods in this book. We have not included them because we feel they deserve a more thorough treatment than can be given in a short section of a book on extension. Instead we indicate which roles these methods can play in developing effective extension programmes and refer readers to more specialized publications on these research methods.

This book is based on one which was first published in Dutch in 1974 (1) and later in German (2) and French (3). Several of the ideas which were first published in these books are included in this edition.

In the third printing of the Second Edition, some small changes have been made and the Guides to Further Reading have been updated.

A.W. van den Ban, Wageningen
H.S. Hawkins, Melbourne

References

1 van den Ban, A.W. (1985) *Inleiding tot de voorlichtingskunde*, 7th edn. Boom-Pers., Meppel and Amsterdam.
2 van den Ban, A.W. and Wehland, W.H. (1984) *Einführung in die Beratung*. Pareys Studientexte 36. Parey, Hamburg.
3 van den Ban, A.W., Hawkins H.S., Brouwers, J.H.A.M. and Boon, C.A.M. (1994) *La vulgarisation rurale en Afrique*. Éditions Karthala, Paris and Le Centre Technique de Coopération Agricole et Rural, Wageningen.

The Authors

A.W. van den Ban (1928) graduated from the Agricultural University at Wageningen, the Netherlands, where he was awarded a PhD in Rural Sociology. He studied also at the University of Wisconsin. For eight years he conducted research on agricultural extension for the Netherlands Ministry of Agriculture, and was later appointed as a Professor at the Agricultural University at Wageningen to start a Department of Extension Education. He was Chairman of this Department for 19 years, and during this period the Department became one of the largest of its kind in the world in terms of student and staff numbers. His former students can be found in many African, Asian and Latin American countries. At present he is a consultant on extension education in less industrialized countries. He has worked in over 20 countries in Africa, Asia and Central Europe, and has received the Excellence in Extension Education Award from the Indian Society for Extension Education.

H.S. Hawkins (1932) graduated in agricultural science from the University of Adelaide, and later completed a PhD in Communication at Michigan State University. He has been Senior Lecturer in Agricultural Extension in the Faculty of Agriculture, Forestry and Horticulture at the University of Melbourne since 1971. In earlier years he spent two years in Argentina as an agronomist, and later served on the staff of the Agrarian University of La Molina in Lima, Peru. More recently he has been involved with Australian university aid projects in Indonesia and with extension training in the Bangladesh sugar industry. His most recent consultancy in 1993 was with the Uruguayan extension service, the Plan Agropecuario.

1 Introduction

Agriculture and agricultural extension are facing a number of serious problems in the present era for which it is not easy to find good solutions.

Food production per capita is decreasing in many African countries, and many do not have sufficient foreign exchange to import more food. In the past much of their increase in food production came from an increased area under cultivation, but this increase is no longer possible, partly because much of the land has decreased productivity due to soil erosion. In many Asian and Latin American countries food production was increasing more rapidly than population growth after the introduction of the Green Revolution. In recent years, however, yields are no longer increasing, although growing populations with increased disposable incomes have caused the demand for food to grow rapidly. For example, to meet this demand in the next decade India will have to increase its cereal production by approximately 30 per cent, or over 50 million tonnes, for which there will be less irrigation water available than is used now. Much farm income in most industrial countries has come from government price support programmes, but this type of support cannot be maintained with the decreasing political power of farmers and international accords such as the General Agreement on Tariffs and Trade (GATT). At the same time agriculture is facing and even causing serious environmental problems.

The former communist countries perhaps face the most difficult problems with the transformation of workers on former cooperative and state farms into private entrepreneurs and with building the infrastructure of input supply, marketing and extension which is necessary to make private farming viable. Everywhere we see a trend towards more competition for world markets in which only the more efficient farmers can survive.

Major changes in world agriculture are necessary before these problems can be solved. Most of these changes require competent farmers who are able to increase productivity, while at the same time maintaining the sustainability of their farming system by making effective use of knowledge and information which is available from or can be generated by several different information sources, such as research institutes, successful farmers and markets.

The more than 500 000 agricultural extension agents in the world have to play a crucial role in increasing farmers' competence. They also are expected to play new roles such as promoting sustainable agriculture for which new skills are required. At the same time their conditions of work are changing rapidly, for example through the privatization of government services, including extension, and the growing role commercial companies and non-governmental organizations (NGOs) play in agricultural extension. Very competent extension agents are needed to make decisions on the future role of the extension services and to implement these decisions. We hope that this book will make some contribution to the development of this competence. Such extension services should be supported by vocational agricultural education for youngsters entering farming.

1.1 Overview

Meanings of the terms 'extension' and 'extension education' vary according to the field in which they are applied. We believe it important for our readers to share our meanings for these terms if we are to develop a coherent and scientific view of extension. Hence, we discuss the origins of extension and dissect the main professional areas in which it is used, paying particular attention to agricultural extension. We outline the main goals or objectives of agricultural extension programmes, and discuss their growing social significance as means for helping rural people to lead better and more productive lives, particularly through the utilization of research results and as an instrument for realizing the agricultural development policy. We point out how extension agents can act as experts who know how to solve farmers' problems, or as guides and mentors who help farmers find their own solutions. We encourage extension agents and organizations to take a systematic approach to their work, basing their plans on a thorough knowledge of farmers they work with, how they perceive their problems and the needs they feel, and to cooperate with various other persons and organizations who generate or communicate knowledge and information which farmers need to make good decisions.

Extension may be used to influence human behaviour. We discuss the advantages and disadvantages of different ways of influencing behaviour. This helps us to understand what we can hope to achieve in extension work and to recognize some limitations or constraints of different extension methods and strategies. Influencing human behaviour is a serious matter which raises many ethical questions. Although we do not prescribe a definite code of ethics for extension agents, we comment on situations and objectives for which we believe extension agents may and should influence people.

Several areas and disciplines in the social sciences have made significant contributions to the development of extension and extension education. These include decision-making, communication, the psychology of learning, of changing attitudes and of perception, and the diffusion and adoption of innovations. We draw from these disciplinary areas to develop a theoretical basis for extension.

Extension is an applied discipline. Hence the practical aspects of application are discussed at some length. We show how the most important methods used in extension are integrated, how their effects are evaluated in planned programmes and how farmers and their organizations can participate in extension programmes.

Most extension agents are associated with large formal organizations such as government departments, universities or commercial firms, although some of the best extension work is done by smaller NGOs. Organizational structure and leadership styles favoured by staff will have a significant effect on extension effectiveness. We also discuss staff development, the roles of specialists and generalists, and the position of female extension workers. The relation between extension work and other development tasks such as regulation and distribution of inputs inside or outside the extension organization is a difficult issue. The Training and Visit System, which has been introduced in many less industrialized countries to increase extension effectiveness, is an example of a comprehensive scheme for structuring the training, delivery and administrative system for an extension service. We discuss these organizational aspects of extension at some length, and conclude the book with observations about the role and duties of an extension agent.

Many of the concepts used in this book are not defined fully in the text but are explained in greater detail in the Glossary. We hope readers will consult this section regularly. Our book is little more than an introduction to a vast field of study which encompasses many disciplines and social skills. We refer to literature related to these wider fields at the end of each chapter and in detailed Chapter notes. Readers wishing to delve deeper should consult the main references for further study.

The main focus of this book is on agricultural extension in less industrialized countries. Farmers in the industrialized countries are businessmen, often with high capital investment in land and machinery, whereas most of their counterparts in less industrialized countries are small-scale commercial farmers or even illiterate subsistence farmers. We discuss some implications of these differences for agricultural extension, but at the same time we believe that many similar principles of change can be applied in both situations. Countries which now are becoming industrialized can learn from the experiences and mistakes of already industrialized countries. When this book is used in a course it will have to be supplemented by materials on agricultural extension from that country.

Many of the principles discussed in this book can be applied to other fields such as health and nutrition education, family planning or extension work with small businesses. We will not discuss these applications, although from time to time we will draw on examples from outside agriculture, especially from health, because it helps to relate theory to the reader's personal experience. For example, it may be meaningful for readers who are smokers to think about their own smoking behaviour when reading that a change in knowledge often fails to change behaviour.

People working in extension programmes may be referred to as extension officers, extension agents and extension workers, depending on the country, the organization and the status of the individual. We have decided to use the term 'agent' throughout this book for two reasons. Firstly, the term 'officer' implies the extension person is a government employee, whereas he or she may conduct extension programmes for an NGO. Secondly, the term 'worker' implies a person of low status in some societies. People working in extension range from high level administrators to field assistants. Hence we have preferred to use a term which does not imply any specific status.

Other terminology has been chosen to suit different parts of the world. We have avoided use of 'developed', 'developing' or 'underdeveloped' countries as these terms imply that all countries should move in the direction of the Western industrialized countries. We believe that an increase in income is desirable for many people in this world, but not for all people, and we are convinced this can be achieved in many different ways. Japan is a good example of a country which now has a higher income per capita than many European countries but has developed its own culture rather than copying a Western model. We write instead about industrialized and less industrialized countries, which we believe are more objective terms to indicate these differences. Such use of these words is not intended to imply that industrialization brings only favourable effects.

1.2 Fundamental ideas on which this book is based

The reader may find it easier to follow the main arguments presented in this book if we outline first the fundamental ideas on which the book is based. These ideas will be explained in greater detail in the following chapters, and some will be the basis of our discussion in several chapters.

We see agricultural extension as one of the policy instruments a government can use to stimulate agricultural development. However, farmers are always free to follow or to ignore the advice of their extension agents. Hence, extension can only achieve changes which the farmers feel are in their interest.

Research findings and other farmers' experiences can help the individual farmer to achieve his own goals more effectively. Therefore agricultural extension must provide an effective linkage between farmers, agricultural research and other sources of information. This makes it possible for extension agents to stimulate learning among their farmers by improving the extent and quality of their knowledge and their decision-making skills. Hence extension agents on the one hand must be competent agriculturalists and on the other hand have learned how to communicate effectively with farmers and how to stimulate learning. This book deals with these latter competencies. We are convinced the extension agent must be oriented primarily towards the farmer's problems as the farmer perceives them, and not towards agricultural technology. The agent has to search for technology which will solve these problems before the farmer can be expected to show interest in learning about the technology.

People can learn from experience. We believe extension agents have a major task in stimulating farmers to learn. Agents themselves also must be eager to improve their work by learning from their own experience, through listening to farmers, experimenting with new approaches and observing and analysing carefully the outcomes of these actions.

Many extension activities involve value judgments on the part of extension agents. At times they must choose between giving information or advice which is in the best personal interests of the farmer and that which is in the best interests of the community at large. Hence it is desirable for extension agents to have considered some of the options open to them for stimulating learning and behaviour change, and to understand the moral and ethical implications of their actions.

Extension involves the reception and interpretation of messages transmitted through different channels. We believe extension agents will be more effective in helping farmers if they understand some of the theoretical background to their work. Hence we discuss ways in which we receive information about the world around us and factors which influence how we interpret the stimuli and messages we receive. We extend these discussions to the ways in which ideas spread in communities and the factors that influence whether the ideas are accepted or rejected.

Extension agents have many different methods available for exchanging information with farmers, research workers and other members of the agricultural system. Depending on their specific goals and the circumstances in which they work, agents may use these methods in different ways to help farmers form sound opinions and make better decisions. The technology for transmitting, storing and retrieving information has developed rapidly in the last fifty years, and the costs of reaching individuals by radio, television and printed matter have reduced considerably in many countries. Hence extension agents must be familiar with the characteristics of the media and how to use them effectively. They also must be aware of the psychological and sociological aspects of group processes and individual extension.

We are convinced that more systematic planning can increase the effectiveness of many extension organizations. This planning requires clear decisions about what kind of changes the extension organization is trying to achieve among what categories of farmers and how these changes will be achieved. This planning must be based on careful analysis of past extension experiences and the reasons why these changes have not been achieved already.

Careful evaluation of objectives, methods and results is an integral part of effective extension planning. Information produced in evaluation is an important tool for extension organization management. The management style should stimulate the learning processes of extension agents.

Guide to further reading

Antholt C.H. (1994) *Getting ready for the twenty-first century: technical change and institutional modernization in agriculture.* World Bank Technical Paper 217, Washington DC.
Antholt C.H. (1994) *Getting ready for the twenty-first century: technical change and institutional modernization in agriculture.* World Bank Technical Paper 217, Washington DC.
IFPRI (1995) *A 2020 Vision for Food, Agriculture and the Environment.* International Food Policy Research Institute and National Geographic Society, Washington D.C.
Tribe, D.E. (1994) *Feeding and Greening the World: The Role of International Agricultural Research.* CAB International, Wallingford.

2 Extension and Extension Education

'When the world is changing very slowly, you don't need much information. But when change is rapid, then there is a premium on information to guide the process of change.' Lester Brown

In this chapter we discuss our meaning for the terms 'extension' and 'extension education', and the role of agricultural extension as a policy instrument for government. The main question to ask when deciding on the objectives of extension programmes is whether such programmes should try to influence farmers to behave in a way which extension agents think is desirable, or whether they should try to increase farmers' capabilities in deciding for themselves the best ways of achieving their goals. Farmers use information from their extension agents, as well as from many other sources. We also give brief descriptions of some professions related to agricultural extension and discuss extension education as a scientific discipline.

2.1 What is extension?

The meaning of the word 'extension' is well known and accepted by people who work in extension organizations and services, but is not well understood in the wider community. As there is no single accepted definition of extension, we will present a number of views and discuss their implications.

Common use of the term 'university extension' or 'extension of the university' was first recorded in Britain in the 1840s. The first practical steps were taken in 1867–68 when James Stuart, Fellow of Trinity College, Cambridge, gave lectures to women's associations and working men's clubs in the north of England. Stuart is often considered to be 'the father of University Extension'. In 1871, Stuart approached the authorities in the University of Cambridge and appealed to them to organize centres for extension lectures under the university's supervision. Cambridge formally adopted the system in 1873, followed by London University in 1876 and Oxford University in 1878. By the 1880s the work was being referred to as the 'extension movement' (1). In this movement the university extended its work to those beyond the campus.

The term 'extension education' has been used in the United States since the beginning of this century to indicate that the target group for university teaching should not be restricted to students on campus but should be extended to people living anywhere in the state. Extension may be seen as a form of *adult education* in which the teachers are staff members of the university. For many years this was mainly an activity of the College of Agriculture which employed 'county agents' all over the state. As the number of farmers has decreased the extension service now tries to serve all citizens with information available from anywhere in the university.

Most anglophone countries now use the American terminology of 'extension', but some have lost the idea that extension is an educational activity. For these the goal of extension is to ensure that increased agricultural production, the major objective of agricultural development policy, is achieved by stimulating farmers to use 'modern' and 'scientific' production technologies developed by research.

2.1.1 Alternative words for 'extension'

We may influence people through extension in many different ways, although people often think only of how it is done in their organization. Therefore, we will illustrate some different possibilities by explaining the meanings of some words which are used instead of 'extension' to describe related processes.

The Dutch use the word *voorlichting*, which means lighting the pathway ahead to help people find their way. Terminology introduced by former colonial administrators often is used in some less industrialized countries, despite the fact that different types of extension might be needed in both countries. For example, Indonesia follows the Dutch example and speaks of lighting the way ahead with a torch (*penyuluhan*), whereas in Malaysia, where a very similar language is spoken, the English and US word for extension translates as *perkembangan*. The British and the Germans talk of advisory work or *Beratung*, which implies that an expert can give advice on the best way to reach your goal, but leaves you with the final responsibility for selecting the way. The Germans also use the word *Aufklärung* (enlightenment) in health education to highlight the importance of learning the values underlying good health, and to stress the point that we must know clearly where we are going. They also speak of *Erziehung* (education), as in the US where it is stressed that the goal of extension is to teach people to solve problems themselves. The Austrians speak of *Förderung* (furthering) or stimulating you to go in a desirable direction, which is rather similar to the Korean term for rural guidance. The French speak of *vulgarisation* which stresses the need to simplify the message for the common man, while the Spanish sometimes use the word *capacitacion* which indicates the intention to improve people's skills, although normally it is used to mean 'training'.

2.1.2 Definition

Our discussion above outlines a number of different processes which have some similarities but important shades of difference. Most agricultural extension services use a mixture of all these processes, but not always the same combination of component elements. The specific mixture used by an organization often is based more on tradition than on a serious consideration of which combination is the most appropriate to their situation. This creates confusion when people from different organizations or trained in different institutions discuss agricultural extension. However, the common meaning for the term is that *extension involves the conscious use of communication of information to help people form sound opinions and make good decisions*. Some differences of opinion about the meaning of extension relate to questions such as:

- In some situations are we concerned only with the formation of opinions, or are we always concerned also with decision-making?
- Should we concentrate solely on increasing farmers' knowledge, or should we also help them to become aware of problems and to clarify their goals?
- To what extent does the extension organization consider one opinion or one decision optimal for its farmer clients?
- Are we satisfied if the farmer makes one good decision, or do we see help with decision-making as a way of teaching him to make similar decisions in the future?

Many extension organizations pay some attention to all these aspects of extension. Agricultural extension agents supply information about agricultural policies and the reasons for them, and endeavour to stimulate certain developments considered to be desirable. For example, they encourage farmers to avoid activities which pollute the environment, and help them to develop into modern and efficient producers. However, the agent's main task is to help farmers with their decision-making.

2.1.3 Providing help and advice

There is strong emphasis in this book on extension as *Beratung*, a process of helping people make decisions by choosing from alternative solutions to their problems. The authors' common backgrounds in agricultural faculties no doubt have influenced this emphasis. We pay attention also to farmers' awareness of their problems and how do they clarify their real goals and opportunities in situations where the extension agent leaves farmers to make their own optimal decisions suited to their personal goals and conditions.

Our earlier definition of extension requires some further explanation. Help implies that farmers' interests are our starting point. Farmers will often decide for themselves what is important, but there are times when the extension agent will decide what is best for them. For example, veterinary specialists may say it is in the best interests of society as a whole, as well as the individual farmer, if certain control or prevention measures are introduced to limit the spread of specific diseases. This problem is discussed in greater detail in Section 2.3.

It becomes necessary to make decisions if the present situation does not correspond with the desired situation. This creates a problem for us only if we do not know how to achieve the desired situation. For example, 133287 divided by 7 is a tricky task for most readers of this book, but not a serious problem. Difficult problems may require outside help such as that given by extension agents if they are to be solved. Simpler tasks can be solved without this help, although a farmer may appreciate help with a task as well. In that case the extension agent provides a service rather than education.

2.1.4 Barriers to farmers

Problems which prevent farmers from achieving their goals are barriers which extension may help overcome, depending on the nature of the problem. Barriers may be grouped as follows:

Knowledge

Some farmers lack adequate knowledge and insight to recognize their problems, to think of a possible solution, or to select the most appropriate solution to achieve their goals. Their knowledge also may be based on incorrect information because of limited experience, upbringing or other cultural factors. The extension agent's objective is to remove the barrier by providing information and insight into the problem. For example, a farmer may not be aware that decline in his or her crop yield is caused by a pest he or she cannot see or identify. The extension agent assists by providing appropriate technical information about the pest and by showing the farmer how to eradicate it.

Motivation

Some farmers lack motivation to behave in a certain way, perhaps because the desired change in behaviour conflicts with other motives. Extension sometimes can overcome this problem by helping farmers to reconsider their motives. For example, they may not pay much attention to milking shed hygiene because disinfectants are expensive and require extra physical effort to use correctly. The extension agent may motivate the farmer to use the recommended methods by demonstrating how shed hygiene improves milk quality and increases financial returns.

Resources

Some extension organizations have the responsibility for removing the barrier of a lack of resources. For example, the extension organizations of ministries of agriculture in less industrialized countries often are responsible for supervising credit and distributing essential inputs such as fertilizers. According to our definition, the organization which provides resources is not engaged in extension, even though it may be doing very worthwhile work.

Insight

Some farmers lack insight into how to obtain the necessary resources. This is similar to the barrier of lack of knowledge and is an appropriate role for extension. For example, the extension agent may provide information about sources of credit, but would not actually grant or supervise repayment of the credit.

Power

It is unlikely that provision of information will bring a change in the power of the farmer and hence this is not a feasible extension activity, except when caused by the following barrier.

Insights into power

Some farmers lack insight into community power relations, or into the power resources they and their colleagues may have at their disposal, and how they could use this power to institute change.

2.1.5 How extension can help

We can see that extension clearly cannot solve all farmers' problems. Appropriate knowledge and insight can be used to solve only some of the problems we have presented, and then only if the extension agents themselves have the required knowledge and insight, or, together with the farmers, can acquire them. Other social functions such as scientific research can help to solve some social problems, for example, by developing methods for increasing crop yields.

Extension agents also must analyse current situations, so that they are in a sound position to give farmers timely warnings of undesirable events or changes in conditions. They also may be able to change farmers' vague feelings of dissatisfaction into concrete problems to be solved. For example, by analysing the economic structure of a farm extension agents may be able to show farmers that they are relying on a particular crop for most of their income, despite declining returns from that crop. Such an analysis also may demonstrate that alternative crops already tested by the farmer have a higher potential for maintaining a satisfactory income level.

Farmers' goals are another important issue for extension agents. Through discussion they can help farmers choose between incompatible goals. Once again we see there is no well defined line of demarcation between extension and adult education.

Now we can define extension systematically as a process which:

- helps farmers to analyse their present and expected future situation;
- helps farmers to become aware of problems which can arise in such an analysis;
- increases knowledge and develops insight into problems, and helps to structure farmers' existing knowledge;
- helps farmers acquire specific knowledge related to certain problem solutions and their consequences so they can act on possible alternatives;
- helps farmers to make a responsible choice which, in their opinion, is optimal for their situation;
- increases farmers' motivation to implement their choices; and
- helps farmers to evaluate and improve their own opinion-forming and decision-making skills.

Extension does not attend to all of these points, nor should it. Given help with one or more aspects of their problems farmers often will be able to solve the rest by themselves. It may be sufficient to define the problem clearly and to analyse it systematically. In other situations it may be enough to provide information which the farmer lacks. Extension agents first must analyse a farmer's situation before deciding the best avenue for providing help. This is discussed further in Section 6.3.1 on mutual discussion.

The term 'extension agent' also may create some problems. Many extension agents do more than give advice. Some supervise regulations or eradicate pests and diseases, while others provide inputs such as fertilizers. Therefore it is unrealistic to say that only those who give advice should call themselves extension agents, but we only speak of extension agents if this is their major task. Extension advice also may be given by people whose main work is outside of extension. For example, rural bank managers may offer advice on different sources of credit.

2.2 Extension and government policy

The previous discussion assumes implicitly that extension agents are free to serve farmers in ways they consider most effective. The reality is that an extension agent nearly always is a member of an organization and is expected to help achieve the goals of this organization, which quite often is a part of the country's Ministry of Agriculture. This limits the freedom extension agents have. Suppose for example, that in a certain area the best way for farmers to increase their income is by growing poppies for heroin, but it is government policy to reduce production of this drug. Extension agents then can expect serious problems with their boss if they teach farmers how to grow this crop more effectively, and even more severe problems if they persuade them to switch to growing this crop.

2.2.1 Government goals

The government will invest in extension when it believes it has value as a policy instrument which helps to achieve government goals. These goals can include the following.

Increasing food production

In many countries the demand for food is increasing because of a growing population, and often because of increasing prosperity. Importation of this food can present a serious threat to the balance of payments, and in large countries would cause a substantial increase in world market prices. If the green revolution had not brought about large yield increases of cereals many people would have been in serious trouble or even have died from hunger.

Stimulating economic growth

This goal is related to the previous one, but it gives more attention to the cost of production and ability to compete in world markets through efficient production. Increased labour productivity in agriculture makes it possible to produce enough food with fewer people. In a country where there is alternative employment for people leaving agriculture, for example in conditions of rapid industrial development, increased labour productivity can contribute to economic growth. This will not happen in a country where people leaving agriculture increase unemployment or under-employment in other sectors of the economy, often leading to serious social problems.

Increasing the welfare of farm families and rural people

Usually farmers and landless rural labourers are among the poorest people in the country. Many governments try, or at least claim that they try, to reduce this poverty because these people have a large proportion of the votes.

Promoting sustainable agriculture

In many countries soil erosion, salination, depletion of aquifers and/or environmental pollution are serious threats to future agricultural production. Hence, many governments show increasing interest in promoting sustainability. These moves may not always be in the short-term interests of farmers or other people affected.

2.2.2 Conflicts

These goals may conflict with each other. For example, increased food production can be achieved in the short-term in ways which are not sustainable, or it may decrease food prices and farm income.

The role of government appears to be changing in many countries. For example, many governments accept less responsibility for development of the country, but place more emphasis on supporting development by the citizens themselves. This change also has implications for the role of the extension service which has less responsibility for modernizing agriculture, but more for enabling farmers to use opportunities opened up by new research findings and/or developments in the market place.

However, there remains a problem that the only power extension agents have to influence farmers is the confidence these farmers have in their extension agents' competence to give good advice and their belief that the agents will give advice which helps the farmers to achieve their goals more effectively. Therefore extension is a suitable policy instrument for problems where the best interests of farmers and extension agents coincide (Fig. 2.1), as was often the case with the introduction of high yielding varieties, but also with many other new technologies.

Extension often is used to achieve government policy goals which are not in the best interests of most farmers. This causes farmers to lose confidence in the extension agents, often making extension ineffective for solving problems where goals of government and the farmers coincide. For example, one country imported much vegetable oil but lacked foreign exchange. Therefore extension agents were ordered to convince farmers to grow more oil crops. This did not have the desired effects because commodity prices made it more profitable to grow other crops. Extension became effective only after the government had changed its price policy, and extension agents had to teach farmers that it was now profitable to grow oil crops and how best to grow these crops.

There are also changes which are in the interest of farmers, but irrelevant to government, or in the interest of government, but irrelevant to farmers. For example, the price of a type of fruit may depend on its colour, but changing this colour is not a part of the government policy. In such a situation extension agents can teach farmers to grow the most profitable colour. Their boss may prefer that they give more attention to helping achieve government policy, but will not object to this teaching.

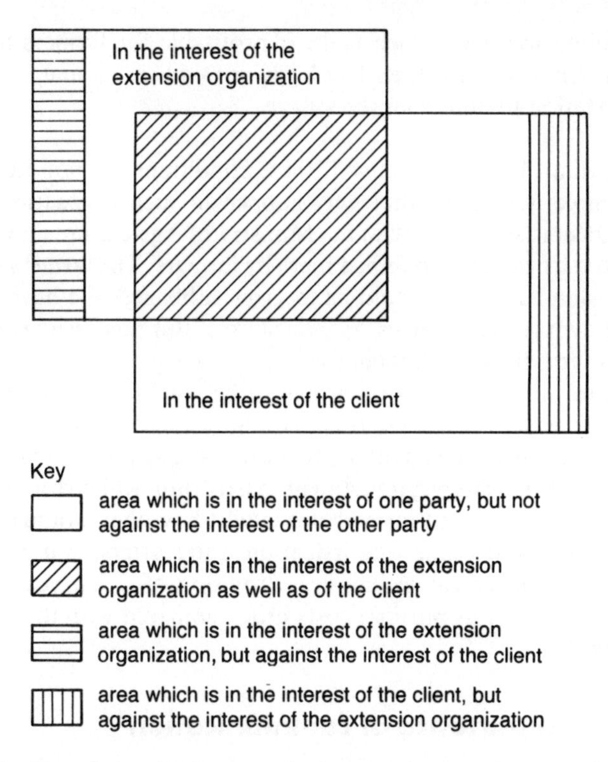

Figure 2.1 The relationship between the interests the extension organization tries to realize and those the client tries to achieve

2.2.3 Returns on agricultural research and extension

Research has shown that investments in agricultural research and extension often give a high rate of return. The average internal rate of return is about 40 per cent which is much higher than for other investments in agricultural development (2). Competent farmers are an important condition for this development, and it is an extension objective to increase this competence. Furthermore, if a farmer is taught a new idea he may use it for many years and stimulate his colleagues to use it also.

However, these studies show there is a large variation in this rate of return, one reason being that extension is not always well organized. One reason is that there are also other conditions for successful agricultural development. In his important book *Getting Agriculture Moving* Mosher (3) mentions the following:

- markets for farm products;
- a constantly changing agricultural technology;
- local availability of supplies and equipment;

- production incentives which make it profitable for farmers to produce more, and not only for their landlords or middleman; and
- transportation to and from the village.

Extension cannot be very effective if these conditions are not met. The following example illustrates what may happen in such a situation: one day when the officers came to work at the Ministry of Agriculture they found the doors blocked by cabbages dumped there by farmers. The farmers told them: 'Your extension agents have told us to grow cabbages. We have been obedient and grown good cabbages, as you can see. But now there is no market for them. So you should sell them for us'.

This implies that extension often has to be combined with other policy instruments to ensure that these other conditions also are met. At the same time other policy instruments often are more successful if they are combined with extension. For example, investment in irrigation will give a much higher rate of return if the rainfed farmers are taught how they can make optimal use of the new opportunities the irrigation water offers, and what are the correct water management techniques. Provision of a few years' intensive extension service for this purpose costs little compared with the costs of the irrigation works.

2.2.4 The importance of communication

Traditionally extension messages are based on farmer experience and/or agricultural research findings. In many countries governmental policies are of increasing importance for decision-making by farmers. We think of price and tax policies, as well as policies related to prevention of soil erosion and environmental pollution and those for regional planning. Usually there are government information departments responsible for communication between government and the public regarding these policies. Their cooperation and division of labour with the agricultural extension service can be organized in different ways depending on national traditions and situations. It is important that it is clear who performs which roles in communication about governmental policies which are important to farmers. Possible roles include:

(1) *Helping* different actors to become aware of problems which require a government policy, and helping them to define these problems as clearly and accurately as possible.
(2) *Analysing* possible solutions for these problems and the consequences which can be expected for each of these solutions. For this purpose it is necessary to know who has information on the causes of these problems and their consequences.
(3) *Deciding* which solution will be preferred. This usually requires nego-

tiation, but someone will have to decide who is allowed to play which role in the negotiation process.

(4) *Informing* the relevant actors about the policy decisions, for example, new rules and regulations, and about the roles they are expected to play in implementing these decisions.

(5) *Monitoring* whether the policies have been implemented according to the ways in which they were planned, and evaluating the extent to which they indeed do solve the problems or to which perhaps they cause new problems.

The communication processes required depend on the nature of the policy making process. Perhaps only a few civil servants and politicians are involved. Increasingly, however, the major stakeholders are allowed to participate in this process. Their information and experience is needed in order to design effective policies, and their cooperation is required for the implementation of these policies. Chamala (1995) has developed a Participatory Action Management Model (4) which can be used to bring these stakeholders together for this negotiation process. In this model decisions at local, regional and national level are interrelated. He uses this model to increase the power of groups who had little influence on these decisions in the past. It is felt that stakeholders who are allowed to participate in the decision-making process will be more inclined to accept that these policies are necessary. This is partly because better decisions will have been made. Therefore they might cooperate in their successful implementation. If most farmers follow the regulations voluntarily, it is possible to force the others to comply, for example, by using fines. It is seldom possible in a democratic country, however, to compel 80 per cent of farmers to follow these regulations.

In several countries increased use of pesticides has caused soil pollution, with a result that consumers will not buy produce grown under these conditions. Therefore the government, farmers' organizations and environmental groups have decided that reduction in pesticide use should be part of the national agricultural policy. Ways of achieving this reduction without lowering agricultural production significantly are decided on the basis of information received from farmers and scientists. More farmers are willing to follow the new environmental regulations voluntarily because their own organizations were involved in identifying the problems and in planning the practical solutions. The farmers now accept that they might lose a share of the market if they do not follow the regulations.

There is a problem with many governmental decisions in that a decision may be quite important for a relatively small number of stakeholders, but only of minor importance for the vast majority. Denying access to the local market to efficient producers of a certain commodity from other countries through tariff protection may be quite important for the well-being of local producers of this commodity. However, it is achieved at the expense of higher prices for local consumers. Government has the responsibility to balance the interests of local producers and consumers even if the consumers are not willing to invest time to participate in this decision-making process. Producers who played an active role in this process may feel that the government did not listen to their arguments if a decision is taken in favour of the consumers.

The extension organization certainly should not assume a policing role in enforcing regulations, because by taking such drastic measures they would lose farmers' confidence, thus rendering the extension agents ineffective.

The extension organization can only perform its roles regarding government policies properly if it has close links with policy makers, as well as with other actors in the Agricultural Knowledge and Information System (see Section 2.4 and Chapter 9).

2.3 Objectives of extension organizations

There are important differences among extension organizations in their objectives. We can use a number of dimensions to analyse these differences (5) (Table 2.1). There is a tendency for an extension organization which is on the left side of one of these dimensions also to be on the left side for the other dimensions, although this is not always the case. We cannot say that a good extension organization should be on the left or on the right side. What is best depends on the objectives and the situation. We will return to these choices in various sections of this book.

Table 2.1 Different attitudes of extension organizations.

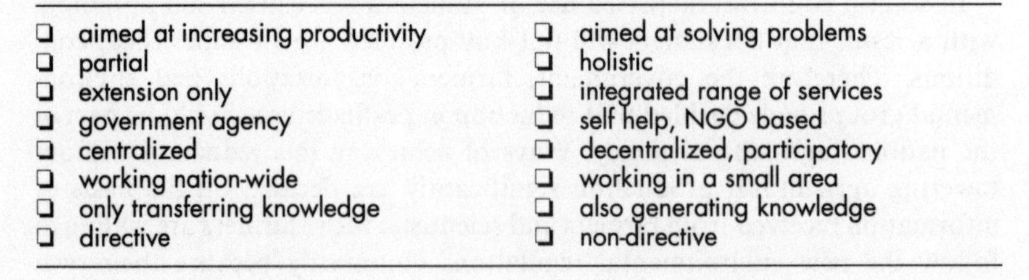

❑ aimed at increasing productivity	❑ aimed at solving problems
❑ partial	❑ holistic
❑ extension only	❑ integrated range of services
❑ government agency	❑ self help, NGO based
❑ centralized	❑ decentralized, participatory
❑ working nation-wide	❑ working in a small area
❑ only transferring knowledge	❑ also generating knowledge
❑ directive	❑ non-directive

The extension service in a country facing a serious food shortage is often expected to minimize the problem by increasing productivity. It is also expected that such a policy will increase farm income at the same time, although it may have more effect on decreasing food prices. This usually implies that the extension administrators believe they know what is good for their farmers and for their country. Hence they order their extension agents to tell farmers what they should do rather than assisting them with finding a solution for their problems. Such an approach is often used by government extension organizations in less industrialized countries by means of the Training and Visit System (see Section 10.9). On the other hand NGOs working in a small area are more likely to adopt a non-directive approach which makes full use of farmers' indigenous knowledge and the results of their experiments (see Section 9.6). NGOs often pay attention to many kinds of problems faced by rural people, not only problems of agricultural production. The government has different agencies for different kinds of problems, which sometimes, but certainly not always, work closely together.

A major distinction among the objectives of an extension service is that extension agents either can try to change the behaviour of their clients in a direction they or their organization considers desirable for the clients, for example to stop smoking, or can help them to decide for themselves which changes are desirable. Some social scientists believe that extension agents should always work in the latter way, whereas many government extension agents are expected by their bosses to convince farmers that they should adopt 'modern' agricultural practices.

We believe that neither way is always right, but that the best way to choose objectives for the extension programme depends on the situation (6). In making this choice consideration should be given to:

(1) Who has the *right* to make this decision?
(2) Who is best informed to make this decision; in other words, who holds the necessary *knowledge*?
(3) What is the impact of the decision-maker's choice on the *motivation* to realize this decision and on the personal development of the farmers?
(4) What is the relation between the extension programme and the agricultural *development policy* of the government?

2.3.1 Right to decision-making

A decision made by a farmer is based partly on his expectation of the consequences of this decision and partly on the way he values these consequences. The farmer's values may be different from those of the extension agent who is often somewhat urbanized, but we see no reason to assume that the values of extension agents and their bosses are better than those of the farmers and their families. Assume, for example, that the use of fertilizers in rainfed agriculture will on average increase family income, but will also increase risk. Then only the family can decide whether income or risk is more important to them. Extension agents might help them to think about the consequences of this decision in a systematic way, but they have no right to persuade farmers to make the same decision they would have made if they were in their position. (See also Chapter 4.)

2.3.2 Knowledge required for decision-making

This does not mean that an extension agent should never persuade villagers to change their behaviour. Many children die from diarrhoea, which could easily be prevented with oral dehydration techniques. Extension agents, parents and children in this situation nearly always will agree about the goals, whereas the extension agent may be the only one who knows how to achieve the goals. Such a situation can also occur in agriculture. Then the extension agent can, and in our view should, use the most effective communication techniques to apply this knowledge.

In agriculture the extension agent often has only part of the knowledge required for decision-making, while the farmer and family have another part. They will know their goals, the amount of capital they have, the labour requirements of their farm in different months, their relations with other farmers, the quality of their land and the opportunities for making money outside agriculture. The extension agent may have some of this knowledge, but usually less than the farm family has themselves. Therefore the combined knowledge of the farm family and the extension agent has to be integrated to develop the most productive farming system for this family. This can be done in a dialogue in which the extension agent listens carefully to the farmer and family, but does not attempt to convince them how they should improve their farming system.

Farmers have been observing the growth of their crops and animals all their lives. Hence they often have very valuable information on agronomy and animal husbandry which research workers and extension agents do not have. This is especially true in rainfed agriculture with its large variation in agro-ecological situations. Farmers might be well informed about the qualities of different plant varieties or about the value of mixed cropping systems for risk reduction (see the reference to the problem solving model in Section 2.5).

Extension agents currently aim more at a sustainable system of agriculture and less at high input agriculture than some years ago. It is generally recognized that indigenous farmers' knowledge is crucially important for developing sustainable agriculture because this way of farming should be adjusted to the local situation which the farmer usually knows better than researchers or extension agents.

2.3.3 Motivation to implement decisions

We do not always bring our good intentions to fruition and neither do farmers. An extension agent might convince farmers to grow vegetables, which will be profitable only if the farmers take good care of them. There is a much higher probability this will happen if the farmers have made the decision to grow vegetables themselves than if the extension agent has made it for them. Persuading farmers to change increases their dependency on government officers, whereas creating a situation in which farmers make good decisions themselves will increase their decision-making skills.

This point is particularly important because village extension workers cannot possibly make good decisions for all their farmers. There is so much variation in farm size, land quality, availability of capital and labour, family goals, etc., that different farmers in the same village often should make different decisions about which production technologies to use.

Development of the human resources among their farmers to make good decisions is an important task for extension agents. The intelligence of their farmers and villagers is a major resource for development which is under-used in many less industrialized countries. Stimulating villagers to make their own decisions helps to develop this resource.

2.3.4 Development policy

As we discussed in Section 2.2, the agricultural extension programme is one of the instruments which a government uses to achieve its agricultural development policy objectives. However, we can only use this instrument effectively if we first learn from the farmers why they do not yet farm in agreement with this policy – why, for example, they do not grow high yielding varieties. It should become an extension programme objective to promote production of high yielding varieties only if their non compliance with the objective is due partly to lack of knowledge or to some extent to lack of motivation. But programme objectives cannot and should not be determined entirely by the agricultural development policy of the government. Awareness of farmers' attitudes should have an important influence on these objectives.

This methodology for choosing extension objectives is relevant for a national extension programme as well as for a dialogue between an extension agent and an individual farmer. In most situations the optimum way to choose objectives is somewhere in between the directive approach, where the objectives are chosen by politicians, bureaucrats or specialists outside the target group, and the non-directive or participatory approach where these objectives are chosen solely by the farmers themselves. It is our view that farmers' skills and potential for making valuable contributions to the choice of objectives are not used sufficiently in many less industrialized countries.

Ways in which a more participatory approach can be applied in planning and implementing an extension programme are discussed in more detail in Chapter 9 'Participation of Farmers in Extension Programmes'. How it can be applied in a small group discussion is discussed in Section 6.2.3. 'Group discussions' and in a dialogue in Section 6.3 'Individual extension'. A participatory relationship between the extension agent and his farmers also requires a participatory style of leadership in the extension organization. This is discussed in more detail in Section 10.3 'Leadership in extension organizations'.

There are a few other points regarding the choice of the objectives of an extension organization which require our attention.

2.3.5 Changing farms or changing farmers?

Most agricultural extension agents have been trained by schools of agriculture how to change farms. They have learned about plant varieties, fertilizers, animal nutrition, etc. However, their role is to change farmers who subsequently may decide to change their farms. Many agents have not been trained in the process of changing farmers, that is in adult education and in communication. They have been taught what to tell farmers, but not how to tell it to them so that the farmers become more capable farm managers. Changing this situation is an important goal of extension education.

Figure 2.2 gives a somewhat simplified picture of the role of agricultural extension. The extension organization obtains information from agricultural research, from agricultural policy decisions and from social and psychological research. This information is used by the management of the extension organization to instruct extension agents what they should tell farmers, in the expectation that such messages will bring about changes in farm management. The main simplification in this figure is that it does not indicate that there is also a flow of information from farmers to extension agents, the managers of extension organizations and the policy makers. We will show in this book how this kind of feedback information is crucially important for successful agricultural extension work and agricultural development policy.

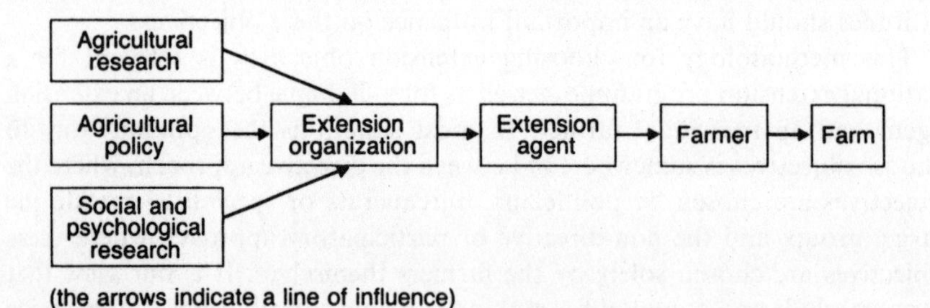

Figure 2.2 Information flow in agricultural extension

2.3.6 Helping farmers to reach their goals

Extension agents can use different ways to help their clients reach their goals:

- giving timely advice to make them aware of a problem;
- increasing the range of alternatives from which they can make a choice;
- informing them about the expected consequences of each alternative;
- helping them to decide which goal is most important;

- helping them to make decisions in a systematic way, either as individuals or as members of a group;
- helping them to learn from experience and from experiments;
- stimulating them to exchange information with colleagues.

2.3.7 Farmers' organizations

Farmers in industrialized countries have organized themselves in many different ways to serve their collective interests. Such organizations play crucial roles in agricultural development in these countries. In less industrialized countries such organizations either do not exist or tend to be ineffective. Establishment of effective farmer organizations is at least as important as the introduction of scientific production technology in many of these countries. Extension organizations can play an important role in teaching farmers how to organize themselves effectively. (See Section 9.6.1 on NGOs.)

2.3.8 Educating farmers

Views about extension agents' roles vary in different parts of the world. They clearly have an educational role in the United States, where extension education is a branch of adult education. Extension agents in most European countries see their role more as one of helping farmers to solve their problems. They are satisfied when the farm becomes more efficient, and are not very interested in changing the farmer. The main extension role in many less industrialized countries is to promote adoption of modern agricultural technology such as use of fertilizers. Yield increase is a major objective in these countries, partly because of the rapidly growing population, and partly because farmers often are considered to be backward and traditional.

We believe that the main way to improve farm efficiency and to increase agricultural production is to educate farmers. However, there are times when it may be better to help farmers solve a problem. For example, they are likely to build only one barn in a lifetime, so it is of little use them learning how to decide which type they should build. In most cases, however, it is must too expensive repeatedly helping farmers to solve the same problem. It will take more time to teach farmers how to interpret a soil test than to tell them which fertilizers to use on the basis of such a test. Nevertheless, repeated discussions of this point after each new soil test take time and hence cost the extension organization money.

Farmers may be educated in two different ways:

- they may be taught how to solve specific problems, or
- they can be taught the process of problem-solving.

For example, you can teach farmers to recognize yellow rust in wheat and to know what to do if their wheat is infected. You also may introduce farmers to a reference book which describes most plant diseases and their treatment, and teach them how to use the book. The latter way initially will take more time and effort by both farmer and extension agent, but may save time in the long-term and also increase the probability that diseases are recognized at the right time and treated in the right way. Hence, this is the preferred way, although it should be recognized that a half-educated person is more dangerous than an uneducated person. Farmers should be assisted to develop good insight into which problems they can solve themselves and which they cannot.

We can see similar developments in general education. Previously students learned many facts, but now they learn much more about where they can find these facts and how they can use them. It has been said that mathematicians know about one per cent of what is known about mathematics at the time they receive their PhD, and perhaps 0.2 per cent of what will be known at the time they retire. Thus it is essential that they learn where to find and how to use the other 99.8 per cent of mathematical knowledge.

Farmers often are eager to learn about new developments. Many farmers in industrialized countries recognize they would be put out of business if they continued to farm in the same way a good farmer did 20 years ago. Hence they must keep well informed about new developments in production technology and markets. We have spoken to farmers in less industrialized countries who also like to learn ways of improving their farms. Extension agents have an important role to play in supporting and facilitating this process of *self-directed learning* (7).

2.3.9 Changing the farmers' situation

For many years we thought that farmers' conservatism was the reason for their failure to adopt new technologies developed for them by agricultural scientists. However, research has shown this view often is not correct, and that the situation in which they farm makes it unprofitable for them to adopt these technologies. For example, it is possible that most of the profits from these technologies go to landlords or to middlemen. Hence, it is hardly surprising that farmers are not very interested in learning about them. If this is the case the farmers will need help to organize themselves effectively in order to promote agricultural development.

Extension agents also may help by reporting the farmers' real situation to policy makers and research workers. This might stimulate them to develop policies which change the farmers' situation, or to develop technologies which are profitable in the present situation. For example, in the 1930s pastures in one part of the Netherlands were rented to the highest bidder for only half the year, thus making it unprofitable for the farmers to invest in pasture improvement through weed control or application of fertilizers. Hence yields also were low. This situation was reported to policy makers by extension agents and farmers' organizations with the result that a new land tenure law was introduced which completely changed the situation, despite opposition from landowners.

2.4 The role of agricultural extension services in the AKIS

The concept of an Agricultural Knowledge and Information System (AKIS) is useful in analysing ways in which farmers increasingly are supported by knowledge and information. It can be defined as

> 'The persons, networks and institutions, and the interfaces and linkages between them, which engage in or manage the generation, transformation, transmission, storage, retrieval, integration, diffusion and utilization of knowledge and information, and which potentially work synergistically to improve the goodness of fit between knowledge and environment, and the technology used in agriculture' (8).

The idea underlying AKIS is that farmers use many different sources to obtain the knowledge and information they need to manage their farms, and that new knowledge is developed not only by research institutes, but also by many different actors. This makes it useful to analyse how all these sources supplement and support each other, or perhaps the kinds of conflicts there are between them. For which kind of knowledge or information does a farmer turn to which source, and how does this source get its knowledge and information? For example, how do agricultural research and extension cooperate? This AKIS also analyses the flow of information from farmers to their colleagues, researchers, policy makers and businessmen.

In order to manage the farm well a farmer needs knowledge and information on different topics, such as:

- findings from research in different production technology and farm management disciplines;
- experiences of other farmers;
- the current situation and likely development in markets for input and products; and
- government policies.

2.4.1 Agricultural research

Agricultural research has made major contributions to increasing agricultural productivity. However, the research does not always provide farmers with the information they need. Research often does not take into account the fact that many farmers, especially in less industrialized countries, have limited access to resources. In one country, for example, much of the research is focusing on questions like: 'How can we increase the potential yield of maize from 7 to 8 tons per ha?', instead of 'How can we increase the average yield from 1.5 to 2.5 tons per ha?' This last kind of research would require cooperation between biological researchers, economists and sociologists to study the access farmers have to different resources and the optimal way they can use these resources (9). These social scientists are scarce in this country and seldom work in a team with biological scientists in Farming Systems Research (FSR).

Another problem is that much of the research is done under more favourable conditions than those found on most farms. For example, the fields on the research institute are ploughed deeper by tractor than the ordinary farmer can do by hand hoe. On-farm trials first have to be carried out under conditions experienced by the average farmer of the area, including, for example, limited supply of labour during certain times of the year, before recommendations can be made which really will increase farm income.

2.4.2 Sources of information

Farmers use many different sources to obtain the knowledge and information they need to manage their farms well. These sources include:

- other farmers;
- government extension organizations;
- private companies selling inputs, offering credit and buying products;
- other government agencies, marketing boards and politicians;
- farmers' organizations and NGOs and their staff members;
- farm journals, radio, television and other mass media;
- private consultants, lawyers and veterinarians.

Few farmers can have direct contact with researchers, especially in less industrialized countries where the number of farmers is large compared with the number of agricultural researchers, the transport system is poor and there is a large social distance between these two groups. Research will only have a real impact on agricultural production if there are others who act as effective communicators between researchers and farmers; we do not say 'who transfer technologies from research institutes to farmers', because communication of the farmers' problems, experiences and situations to the researchers is at least as important in order to obtain research findings which are relevant and really important for agricultural development. Furthermore, development of technologies which are profitable in a certain situation requires the integration of knowledge from different sources such as research and the market place. Frequently much of this integration is left to the farmers, although extension agents and researchers also may play a role in the process. Integration is an important part of the work of FSR who try to bring together research information from different disciplines with information from farmers' experiences, and sometimes also with information on market developments and government policies. Commodity researchers do not always appreciate that their information has to be combined with information from other sources before it becomes valuable for supporting farmers' decision-making.

In Section 5.6 'Adoption and diffusion of innovations' we discuss the importance of the various information sources for farmers in more detail. We show there that farmers usually learn most from their colleagues. In villages with a rapidly modernizing agriculture the most influential farmers often are those who have experimented with new farm practices and adapted them to the local situation. In industrialized countries the role of commercial companies in providing knowledge and information is increasing. It is quite common for such companies to employ more extension agents than the government extension organization, although part of their work may include sales. For example, most of the dairy extension specialists employed by government services in Australia in the 1980s now work for commercial companies and cooperatives. These companies can make more profit if the products they sell are used in well managed production systems, and the products they buy from farmers are of good quality. An increasing proportion of the agricultural research also is done by private companies selling seeds, fertilizers, pesticides, farm machinery, etc.

In Section 5.6.3 we also discuss how innovations usually have a hardware and a software aspect and that information about the hardware generally is developed by different actors from those who produce software information. Farmers may play a major role in developing the latter kind of information.

2.4.3 Researcher and extension agent working together

An effective linkage between agricultural research and extension is quite important for the development of agriculture, but in many countries this linkage is weak. Often this weak linkage is caused by researchers who are rewarded more for results which are appreciated by other researchers than for results which make a real contribution to solving farmers' problems. For example, publications in scientific journals are often a condition for promotion in a research organization, whereas time spent cooperating with extension agents in developing recommendations may not be rewarded. This time cannot be spent on writing for other scientists.

An effective linkage requires researchers and extension agents who visit farmers together to analyse their problems and to discuss which solutions are possible. Development of good solutions requires contributions from researchers who have a deep knowledge of certain aspects of the problem, and from extension officers who should have a broad knowledge and understanding of the farming system. Many extension officers in some countries do not in fact have this broad knowledge and understanding, because lack of transport and/or paperwork too often confine them to the office.

This joint analysis of farmers' problems can be done by FSR and by Rapid

Rural Appraisal (RRA). In RRA a team of researchers of different disciplines analyse a cross section of the farms in an area to discover what are the major problems faced by farmers and what are the reasons farmers farm in the way they do. Recently RRA has been criticized as being part of a top-down approach. Now farmers themselves often are involved in this type of research which is called Participatory Rural Appraisal. There are many publications which describe these research methods in more detail, such as Pretty *et al.* (1995), Chambers (1991) and Mettrick (1993) (10).

It is also desirable that on-farm trials are a joint responsibility of researchers and extension agents. For decisions about what kind of trials are most needed, researchers know most about possible outcomes of these trials based on research results at their institutes and their knowledge of the theory. On the other hand good extension agents know more about which information farmers lack for good decision-making, and what farmers' experiences have been with previous research recommendations.

Extension agents are in a better position to supervise on-farm trials closely and to make the necessary observations, whereas researchers are well placed to plan good research designs and to analyse the data properly. Extension agents also are more motivated to disseminate findings of investigations in which they have participated than findings from trials reported in the scientific literature. Hence, it is important for researchers and extension agents to work together as a team on the development of agriculture in their area. This cannot be achieved if researchers look down on extension agents, who usually have a lower level of education.

2.4.4 Farmers' knowledge

Much agricultural knowledge is developed in the course of simple experiments that many farmers make, such as when they grow a crop which is new to the area or when they change the date of fertilizer application. They also learn from their attempts to adjust extension recommendations to the situation on their farm. Researchers and extension agents can cooperate by helping farmers to make these experiments as good as possible and by assisting them with drawing the correct conclusions from the results. This cooperation will increase the quality of the information available and decreases the probability that farmers will refuse to follow a recommendation. They might learn from their own experience that an idea or a technique will not work, because they made a mistake in their experiment. These experiments by farmers usually generate information about the need for labour and capital in different seasons for new technologies, and the availability of these resources. Such information is crucial for developing appropriate technologies.

The valuable knowledge gathered by farmers over generations, so called indigenous knowledge, is often neglected by researchers, although this information can be quite important for location-specific recommendations and for developing sustainable farming systems (11).

Modernizing agriculture often requires a change in the farming system, for example by growing different crops or by integrating crop and animal production in a different way. It is not feasible on a research station to study what the consequences of these changes will be in the actual farm situation, such as demand for labour in different periods. This kind of information is usually generated by innovative farmers who might cooperate with researchers and extension agents in this process. Extension agents can learn from these farmers much of the information they require in order to provide other farmers with good quality help.

2.4.5 Analysis of the AKIS

It is important to analyse the AKIS for a certain region or a certain branch of agriculture in order to discover possible gaps which hamper agricultural development, as well as possible overlaps which waste resources and might cause conflicts. For example, if policy makers are not well informed about the actual situation on farms, they are not able to design effective policies and extension programmes. If feed companies give good advice about ways to prevent poultry diseases, there is little reason why the government extension organization should spend taxpayers' money to perform the same task. This type of AKIS analysis can be the responsibility of the extension service. It should be used to improve the coordination between the actors in the AKIS, for example by discussing how each organization can contribute as much as possible to make the AKIS more effective. Farmers' organizations also can play a valuable role in the process by formulating the information needs of their members and stimulating research institutes, extension services and other actors to provide this information.

Engel (1995) developed a methodology to involve the actors themselves in the analysis of their AKIS: the Rapid Appraisal of Agricultural Knowledge Systems. This can be the starting point for a cooperative process to make this AKIS more effective (12).

Agricultural information can be incorporated in technological products such as pesticides, machines or seeds and the accompanying steps for their use. These are called *technologies* by Bennett (1990). Other information and skills are not incorporated in a product, such as scouting for insect invasions, time of weeding or crop rotations. These are called *practices*. Commercial companies can often make a profit by selling and providing information about technologies, but information about practices will only be spread by extension services funded by taxpayers or farmers. Various technologies and practices can also be combined in a *production system*, such as a system for producing tomatoes or for low input agriculture.

It is especially important to conduct such an analysis of the AKIS before starting a new project. Unfortunately, donor funded projects often do not strengthen the existing system, but develop a new organization which overlaps with other elements of the system and tends to collapse after the donor aid expires (13).

When analysing the AKIS attention should also be given to the fact that different groups of farmers, such as male and female farmers, crop and livestock farmers, and farmers with different levels of resources, possess and require different kinds of information and use different information sources.

2.4.6 The role of the AKIS

The traditional role of the extension organization in less industrialized countries is the transfer of technologies (TOT) developed at the research institutes to farmers. A major role in industrialized countries has always been to learn from the experience of the most successful farmers in order to teach other farmers how they can improve their farm management. Quite often analysis of the AKIS shows that a role other than TOT is more appropriate. This could involve providing farmers with a basket of opportunities and helping them to choose the right opportunity for their situation.

Another role of the extension organization can be to help farmers:

- to experiment with new technologies or with new farming systems;
- to gain access to relevant information from a variety of information sources;
- to evaluate and interpret this information for their own situation; and
- to learn from their experiences.

Development of a network to exchange information among all relevant actors is an important aspect here. McDermott (14) has shown that it is usually necessary to integrate the information from researchers, farmers and extension agents to be able to develop technologies which work well in a given situation. This process of integration often receives insufficient attention. In the past there was a clear distinction between researchers, extension agents and farmers. Gradually this distinction becomes blurred as they participate together in a learning process to find ways of improving farming in which each group learns from all other participants in the process.

It becomes necessary to form a network of different actors in which the information each of them has can be integrated. This networking role can be performed with information on production technologies, as well as for developing more effective relations with the world outside the farm, especially the markets.

2.4.7 The TOT approach

The starting point in the TOT approach is a new technology developed at a research institute. This is different from modern extension where, as a rule, the problem farmers face is taken as the starting point. The extension agent's role is to help farmers to discover, develop and evaluate relevant information for solving the problem. This often will include information on new technologies developed at research institutes. It also will include information from many other sources which has to be integrated with the research information to develop a solution which helps the farmers to achieve their goals more efficiently.

We do not deny that TOT is sometimes desirable, but this depends on the problems farmers face and on the technologies available. TOT can be the best approach when a new plant variety becomes available which is resistant to a major plant disease, but which does not require different cultivation practices from those of the varieties currently in use. However, extension quite often would have been more successful if TOT had been replaced by an approach which integrates information from different information sources, including the farmers themselves.

Agricultural development will be more successful if the different actors in the AKIS share their knowledge and stimulate each other to develop new ideas about the direction in which agriculture should develop, and on the way this development can be achieved.

2.5 Models of research-extension linkage

Agricultural research findings are of little use if they are not adopted by farmers. Unfortunately, the only effect of much research is to produce an interesting article in a scientific journal or a report which lodges on library shelves because there is no effective research-extension linkage.

Havelock (15) has developed three models which are useful to analyse how this linkage works.

2.5.1 The research, development and diffusion model

Havelock's research, development and diffusion model is often used in industry. It speaks of:

- basic research
- applied research
- development
- diffusion.

Development of the now common fluorescent lighting tubes required basic research into the processes of electrical discharge in certain gases, applied research on management of these processes and development of efficient production techniques. Diffusion is the task of the marketing division, which also has considerable influence on which of the many possibilities thrown up by research is actually developed. They know which products can be sold.

Diffusion is not a real problem for some industrial products. The fluorescent light did not require an extensive communication and educational programme to teach people that it uses less electricity than the traditional bulb to produce a comparable intensity of light. However, about 90 per cent of innovations developed by industry fail to attract enough buyers to make production profitable.

2.5.2 The social interaction model

A second model, the social interaction model, stresses the diffusion of innovations. This shows that mass media plays an important role in creating awareness that these innovations exist. However, most people wait until they discuss the innovation with others who already have experience with it before adopting the innovation themselves. The model assumes that innovations have been developed which are profitable for large numbers of people. Research among farmers has played an important role in development of this model, which we will discuss further in Section 5.6.

2.5.3 The problem solving model

A third model, the problem solving model, starts with the person who has a problem rather than with research or the innovation. Farmers will need information to solve their problems. Part of the information will come either from the existing stock of research findings or from new research carried out to solve this problem. Other information will come from farmers and their families, especially that relating to their experience and their goals or those of other farmers. A farmer's son might join his father after leaving school. The farm then will have to be enlarged to provide income for two people through purchase of more land or by intensifying land use. Research will give some indication of the income and risk to be expected from different alternatives. However, it will not provide information about the farm family's interest in different enterprises or the consequences for their relations with others members of their community. This model is not very helpful when analysing how we can reach a large number of farmers who face similar problems.

2.5.4 Applying the models

We cannot say that one of these models is right and the others are wrong – each is useful for analysing problems we face under certain circumstances.

An extension organization can play an important role in the latter two models. The major role in the second will be to diffuse research findings among the farmers. The major role in the third model is to help farmers clarify exactly what their problem is and to find or to develop the information required to solve it. In the diffusion model extension agents act as experts who teach farmers new knowledge, while in the problem solving model they act more as guides and mentors who help farmers decide which way they would like to go. Extension agents should learn which of these roles they should perform and when they should perform it. Ideally, they should be able to perform both roles, and select the more appropriate role or combination of roles according to circumstances.

The social interaction model has been used less frequently in recent years, while use of the problem solving model has increased. This is because we are more aware of the undesirable consequences of many innovations, and the fact that farmers' knowledge and experience is a very important resource for developing good solutions to their problems. Some people may have gone too far in this direction, and have underestimated the contributions agricultural research can make.

Most extension organizations in less industrialized countries follow the social interaction model, but they may have to think more often in terms of the problem solving model. This is because only the farmers and their families can decide the best solution to many of their problems, especially where values are involved, such as how much money to borrow for increasing farm production. Furthermore, part of the information required for solving an agricultural problem often has to come from farmers themselves, as we have discussed already in Section 2.3.

2.6 Professions related to agricultural extension

Agricultural extension organizations try to change farmers' behaviour through education and communication. Many other professions also try to change people's behaviour through one or other of these means. Agricultural extension often has worked in isolation from these other professions, whereas much could be learned from their experiences and research, even though they face somewhat different problems. It can also be in farmers' best interests if extension organizations cooperate with these professions. Therefore we will discuss briefly these professions and the problems they face.

2.6.1 Health, nutrition and family planning

Health and nutrition education stimulate people to behave in such a way that many diseases can be prevented. Their task is somewhat more difficult than that of agricultural extension because the relationship between cause and effect is more difficult to detect, health problems being related to death are emotionally highly loaded, and usually less staff are available. Close cooperation between agricultural and health educators is highly desirable as health problems are a major hazard for many farm families (16). The same is true of family planning education, a field in which much valuable research has been done. Some family planning programmes have not been very effective because they stress that planning is in the collective rather than the family interest (17).

2.6.2 Home economics

Home economics extension work in the United States is closely related to agricultural extension because of the many close ties between farm and home. Some less industrialized countries have followed the American example by giving female extension agents home economics rather than agricultural tasks. This might not be a good decision in a country where most of the agricultural work is done by women. Also, home economics is important for agricultural development in these countries. For example, the time needed for fetching water and collecting firewood cannot be spent on agricultural production.

2.6.3 Environmental issues

Extension work related to environmental issues and nature conservation is growing rapidly. The fact that much more attention is paid to these issues now than 20 years ago is partly the result of work conducted by groups of concerned citizens, although they might have achieved more if they have been more target group oriented, and had tended to preach less to the convinced.

2.6.4 Change of occupation

Society is changing so rapidly that many of us will have to change occupation several times during our life. Hence career guidance has become more important. Guidance officers pay more attention to counselling techniques (see Section 6.3) because it is important for people to get a clear picture of their own goals before making a decision. It is clear that many farmers also will have to change their occupation.

2.6.5 Education

People are showing more interest in the diffusion of educational innovations because they are convinced that an effective educational system is important for the future of their country. Goals of change programmes are more difficult to define here than in agriculture. Research on this topic conducted by Havelock and others has considerable significance for extension (18).

Some of the innovations introduced in schools can be modified for use in extension education. It has been recognized increasingly that education is not confined to youth. Hence continuing education is regarded as a necessary process. Much of the education we received in agriculture and in other fields will be out of date well before we retire. In agriculture we can learn much from the adult education techniques used in industry. For example, adult educators place considerable emphasis on training in interpersonal relations. This is also an important area for agriculture because of the increasing contact between farmers and the outside world.

2.6.6 Organization development

Organization development (OD) also has attracted increasing interest in recent years. Consultants using OD techniques help organizations to become more effective, a useful possibility for many extension organizations. A strong point of OD activities is creation of situations in which people discover for themselves that they have to change (19).

2.6.7 Government information services

Government information services inform the public about government policies and explain these policies. They also give citizens a means of influencing government policies, such as zoning in the area in which they live. Some government information officers face a dilemma over whether they should promote acceptance of government policies or help citizens develop their own informed opinions about these policies.

2.6.8 Other professions

Agricultural extension also can borrow ideas from journalism, audio-visual communication, advertising, political and religious propaganda.

2.7 Extension education

Extension education deals with strategic questions associated with the extension process. It collects and integrates, where possible, existing knowledge about this process from other scientific disciplines, and adds to knowledge through extension research. Extension education can help extension managers and agents make rational decisions about the goals, methods and organization of extension, taking into account the relationships between these goals and methods. Extension agents wishing to give sound advice to farmers must not only understand the extension process but also have adequate technical knowledge of the discipline in which they give advice. For example, they must have a thorough knowledge of animal husbandry if they are to be livestock extension agents. Similarly, a chemistry teacher must be both a chemist and an educator.

Extension education tries to answer strategic questions in order to help extension organizations and agents carry out their work. For example:

(1) How and when do you decide to give or withhold extension advice?
(2) What is the goal or objective of the extension advice? Do extension agents give farmers solutions to their problems or do they teach them how to solve their own problems in the future? How does the extension organization combine Ministry goals for agricultural development with farmers' goals?
(3) To whom is extension advice directed? Is it directed to everyone living in a certain area, to specific groups, or to those about to make a decision? Unfortunately, those who, by certain standards, most need extension advice seldom ask for it.
(4) Should the extension organization aim to help farmers solve all their problems, or should help be restricted to one type of problem? For example, should the organization try to help villagers with all their needs, or should it work only with the problems of a single crop? Extension agents who concentrate on a single crop have the advantage of developing greater expertise. On the other hand, they may not be able to solve other related problems which farmers consider important, or to see the relationships between this crop and their other crops and animals. Furthermore, the extension organization may be able to provide only a few extension agents for this specialized problem who are based a long way from their prospective farmer clients.
(5) How do extension agents decide which is the first problem for them to investigate? For example, livestock extension specialists must decide whether they will concentrate first on genetic improvement or on better nutrition.
(6) Do extension agents attempt to give farmers a solution to their problems or do they try to guide farmers to solve the problem themselves? (See Section 2.3)
(7) How does an extension organization or agent develop a solution to the farmer's problem? Can they use existing knowledge, either from research results or from the experience of successful farmers? Should they first stimulate new research or local verification of research done elsewhere? Can they develop a solution themselves or with the help of some farmers by studying research done in different disciplines and/or drawing on farmers' experience? Should this solution be tested first?
(8) Which combination of extension methods should be used? Is this

combination based on a systematic analysis of the farmer's situation and on the goals of the extension programme? Which method does the extension agent use to achieve different sub-goals, including stimulation of interest in the programme, influencing group norms about the problem, or helping farmers with their final decision?

(9) How are each of these methods applied in practice? How do extension agents apply the methods to specific goals they hope to achieve, and to development of the farmers' knowledge of and attitude towards the problem?

(10) How is the extension service organized? (See Chapter 10.)

(11) With which other organizations will the extension service cooperate, and how can this cooperation be organized? (See Section 2.4.)

(12) How effective is the linkage between research and extension? Which factors influence this effectiveness?

(13) How do we evaluate the work of an extension organization? (See Chapter 8.)

(14) How big should the extension budget be and how should it be apportioned? For example, what proportion of the budget for irrigation works should be used to teach farmers how to use irrigation water correctly? What proportion of this budget should be used in mass media programmes, group meetings or on personal visits?

Most extension research until recently has been directed towards answering Questions 8 and 9 relating to choice and use of various extension methods. More research interest now is being shown in Questions 2, 6 and 11 which are concerned with extension goals, the relations between extension agents and farmers and the AKIS.

Our discussions above may have implied that the initiative for giving extension advice lies with the extension agent. In fact the initiative often comes from farmers who recognize the need for expert information for their decision-making and turn to the extension agent for help. Many extension agents do little more than try to satisfy these demands, which has the advantage that their work then coincides with problems farmers consider to be important. The disadvantage is that extension agents cannot help farmers become aware of problems they have not thought of. From this it follows that extension cannot reach some groups in the population very easily because they have not yet identified any problems. Thus there is the danger that extension agents will help to solve yesterday's problems but not those of tomorrow.

2.7.1 Extension education as a science

Extension education as a science differs somewhat from many other sciences. Most scientists try to explain why things are as they are. They realize that information can be used to change things, but they do not worry very much whether this happens or not. Thus, we can describe scientific disciplines like physics or biology as being conclusion-oriented.

Other scientists try to find applications for their explanations such as to improve soil fertility. This also is true of some research in extension education, for example, research on the extent to which an extension organization reaches different categories of farmers. Applied scientific effort often ceases when scientists have produced findings which can be applied. Many scientists regard application of their findings as someone else's responsibility.

Much extension education research goes one step further by regarding the research itself as a tool to achieve change. Information gained from studying how processes of change take place is used to influence these processes. An extension education research worker might first study which decision has to be made by extension agents or managers, what information is required for this decision and how much of this information already is available. Then the worker will try to provide the information which is still lacking. We can call this a decision-oriented science. It may help extension agents and farmers to become more aware of the problems they face. We often make decisions unconsciously. For example, extension managers might be so busy with all kinds of bureaucratic problems that they do not think very systematically about the possibility of conflicting interests between different groups of farmers and the role their staff play in these conflicts. A research worker asking questions about these conflicts might stimulate the manager to make decisions about the role of the organization. In this way the extension researcher may not provide solutions to problems of extension agents or managers, but may guide a systematic process in which they find these solutions themselves, perhaps together with their farmers.

There is an important difference between the physical and social sciences. Measurement in the physical sciences seldom will change the object being measured. On the other hand, human beings we study in the social sciences may well change as a result of measurement.

Decision-oriented research which intervenes in a social process often will give deeper insight into the way a society functions than conclusion-oriented research. Studies in power relationships in a village community are a good example. We gain a clear insight into these relationships by attempting to introduce change in a community and then observing and analysing the change processes. An extension agent who understands the relationships between class, kinship and power in a community will be in a sounder position to introduce what he or she considers to be desirable changes.

Extension education builds on several conclusion-oriented sciences to help extension agents with their decisions, especially psychology, sociology and anthropology. Other sciences which contribute to better extension work also can be used. We find much the same situation in medicine which not only builds on biology, but also is willing to use aspects of psychology, electronics or any other discipline which may contribute to improved patient care. The main limitation clearly is that none of us can master all aspects of science which can make a useful contribution to agricultural extension.

2.8 Chapter summary

Extension education involves the conscious use of communication of information to help people form sound opinions and make sound decisions. Usually this has an educational objective also: to learn to form opinions and to make decisions. It can be an effective policy instrument for stimulating agricultural development in situations where farmers are unable to reach their goals because they lack knowledge and insight. It should be used as a policy instrument, only if it is in the best interests of government or the organization which pays for the extension service, for the farmers to achieve their goals.

An extension organization can try to achieve changes in a direction which it considers to be desirable for the farmers, such as better control of plant diseases. It also can help farmers to achieve their own goals more successfully, for example, in choosing between a farming system with high average income and high risk, and a system with low income and low risk. Choosing their own goals should be preferred when values play an important role in decision-making.

Extension is an effective instrument only when combined with others, such as research, provision of inputs and credit and marketing. It can teach farmers how to produce crops and animals in the most profitable way, as well as how to organize themselves in cooperatives and other farmer organizations.

Farmers get information from many different sources in the AKIS. This information should be integrated for decision-making, either by farmers, by extension agents or by others. Farmers are not only receivers of information, but they also develop information themselves for their own use and for sharing with their colleagues.

The return to investments in agricultural extension often are high when extension and research are well organized and coordinated.

Agricultural extension has much to learn from experiences with changing human behaviour in other areas such as health education. Extension education is a decision-oriented science which applies conclusion-oriented social sciences. It supports the strategic decisions which have to be made in an extension organization.

☞ Discussion questions

1 Analyse a situation in which you have received extension help or advice. What do you think were the extension agent's goals when he gave you this help? To what extent did they agree with your own goals? Does our definition of extension coincide with the extension agent's task in this situation? Could you suggest a more appropriate definition of extension?

2 What aspects of extension coincide with your own expectations when you started reading this book? In which respects do you think that a different view of extension would have been a better starting point? What are the consequences of these points for studying this book?

3 Who are the major actors in the AKIS in your country? What kind of information does each actor contribute? Is there a system of coordination and cooperation between (some of) these actors?

4 What can you achieve with extension programmes over the use of herbicides, pesticides and other dangerous chemicals? What factors are likely to limit your success?

5 Research has developed a method to increase maize yields in country A by using hybrid seeds, fertilizers and pesticides. Most farmers will have to borrow money to purchase these inputs. Usually they will earn a high rate of return on this money, but in a very dry year they may lose the lot. With the population increasing rapidly, the demand for maize also is rising fast. Which strategy should the extension service use?

(a) Stimulate the farmers to use this new technology, or
(b) help farmers decide for themselves whether or not to use this technology.

Why?

6 In section 2.2 we discussed the point that agricultural extension can be used as a policy instrument to help farmers reach their own goals, and as an instrument to reach government goals. Think of examples:

(a) where both policy objectives make sense to you;
(b) where agricultural extension has been used to reach government goals, but where this was not the right policy instrument, or where it should have been combined with other policy instruments;
(c) where the government used agricultural extension as the only policy instrument to reach their goal, but where they should have used it in conjunction with other instruments.

7 Is the extension organization in your country expected to transfer technology from the research institutes to the farmers? Do you think that this is a major role this organization should perform? Why?

Guide to further reading

The oldest textbook we know of on agricultural extension is:

Tschajanow, Alexander (1924) *Die Sozialagronomie, ihre Grundgedanken und Arbeitsmethoden*. Paul Parey, Berlin. (Originally published in Moscow, 1917.)

There are Chinese books which are much older but we cannot read them.

Other references include:

Albrecht, H., Bergmann, H., Diederich, G., Grosser, E., Hoffmann, V., Keller, P., Payr, G. and Sülzer, R. (1990) *Agricultural Extension*. Deutsche Gesellschaft für Technische Zusammenarbeit (GTZ), Eschborn.

Bennett, C. (1990) *Cooperative Extension Roles and Relationships for a New Era*. Extension Service, US Department of Agriculture, Washington DC.

Chambers, R. (1997) *Whose Reality Counts? Putting the Last First*. Intermediate Technology Publications, London.

Havelock, R.G. (1969) *Planning for Innovation Through the Dissemination and Utilization of Knowledge*. University of Michigan, Center for Research on the Utilization of Scientific Knowledge, Ann Arbor, Michigan.

Merrill-Sands, D. and Kaimowitz, D. (1991) *The Technology Triangle; Linking Farmers, Technology Transfer Agents and Agricultural Researchers*. ISNAR, The Hague.

Mosher A.T. (1966) *Getting Agriculture Moving: Essentials for Development and Modernization*. Praeger, New York.

Röling, N.G. (1988) *Extension Science; Information Systems in Agricultural Development*. Cambridge University Press, Cambridge.

Röling, N.G. (1994) Agricultural knowledge and information systems. In: *Extension Handbook*, (ed. D.J. Blackburn) 2nd edn. Thompson Educational Publishing, Toronto, Ch. 7.

Scoones, I. and Thompson, J. (eds) (1994) *Beyond Farmer First; Rural People's Knowledge, Agricultural Research and Extension Practice*. Intermediate Technology Publications, London.

Swanson B,E. (ed.) (In press) *Improving Agricultural Extension*. FAO, Rome.

3 Methods of Influencing Human Behaviour

In the previous chapter we have shown that extension offers limited opportunities for changing the behaviour of our fellow man. Governments wishing to change some aspect of the behaviour of the population often will use policy measures such as laws or subsidies rather than extension or information programmes, although they may combine these measures with extension programmes. We need to understand the different methods which can be used to influence human behaviour if we want to know when it is feasible or desirable to use extension. Some of these methods are discussed in this chapter. They are not mutually exclusive, often merging one into another. The methods to be used will depend partly on the values of the observer.

3.1 Compulsion or coercion

Power is exerted by an authority, forcing somebody to do something. People applying coercive power require the following conditions:

- they must have sufficient power;
- they must know how they can achieve their goals; and
- they must be able to check whether the person being coerced is behaving in the desired manner.

Application of coercive power means that the people applying the power are responsible for the behaviour of the people they are trying to change. It is possible to achieve behavioural change with a large number of people in a relatively short time using this method. However, it can be very expensive to maintain and control, and the people being coerced may not always behave as required. The method is unsuitable for changing behaviour which requires initiative by the people being coerced. Extension may be essential to make the sanctions known, and to try to persuade the people being coerced to follow regulations of their own free will. Many government regulations and laws relating to public health, traffic control, etc. are of this type. Governments use this method in an attempt to prevent farmers from polluting the soil or water, or from causing soil erosion. People are likely to resume their old behaviour as soon as coercion ceases.

3.2 Exchange

Goods or services are exchanged between two individuals or groups. The conditions necessary for applying this method are that:

- each party in the exchange process considers the transaction to be in their favour;
- each has the goods or services desired by the other; and
- each can only deliver his part when the exchange goods or services have been delivered by the other, or he can trust the other to comply.

Exchange often is a very efficient method for meeting the needs and interests of different groups, parties or individuals. However, it is not always efficient or fair. Sometimes the other party is inclined to deliver as little as possible of the expected exchange. We see this situation in industrial negotiations between employers and employees, and in trade negotiations between peasant farmers and city merchants. Extension can play a useful role by drawing the attention of a potentially disadvantaged partner in an exchange to ways of preventing the other partner from gaining an unfair advantage. For example, farmers in a remote part of a less industrialized country can be given information about the prices paid for the produce in urban markets. They also can be given advice about ways of ensuring fair and legal trading arrangements with their urban-based trading partners.

3.3 Advice

Advice is given on which solution to choose for a certain problem. We can use this method if:

- farmers agree with us about the nature of their problems and the criteria for choosing a 'correct' solution;
- we know enough about the farmers' situation and have adequate information to solve their problems in a way which has been tested scientifically or in practice;
- farmers are confident we can and will help them with a solution to their problems;

- we do not think it necessary or possible for farmers to solve the problem themselves; and
- farmers have sufficient means at their disposal to carry out the advice.

Advisers are responsible for the quality of their advice. While the advisers' specialized knowledge may be put to good use, there is usually little development of the farmers' capacity to solve their own problems. Doctor–patient relationships and many advisory situations between extension agents and farmers are good examples of this method.

3.4 Openly influencing a farmer's knowledge and attitudes

This method may be applied when:

- we believe farmers cannot solve their own problems because they have insufficient or incorrect knowledge, and/or because their attitudes do not match their goals;
- we consider that farmers can solve their own problems if they have more knowledge or have changed their attitudes;
- we are prepared to help farmers collect more and better knowledge to help them change their attitudes;
- we have this knowledge or know how to get it;
- we can use teaching methods to transmit this knowledge or to influence the farmers' attitudes;
- farmers trust our expertise and motives, and are prepared to co-operate with us in our task of changing their knowledge or attitudes.

It is possible to achieve long-term behavioural change using this method. The farmers' self-confidence and capacity to solve other similar problems by themselves in the future are increased. It is a labour intensive method which often is used in extension and education programmes. For example, extension agents may teach farmers how to control insect pests in their crops with the strategic use of pesticide sprays. Their first task will be to explain the life-cycles of the insect pest and the crop so the farmers understand when each are most vulnerable to attack. If the farmers know and understand how to use pesticide sprays safely and at the most vulnerable times of the insect's life-cycle, they will be in a better position to solve similar problems in the future. This means they are less likely to ask extension agents for advice whenever there is an insect problem, but will use their knowledge and experience to solve the problem themselves.

It is possible also to try to influence only knowledge level or only attitudes. Most of the conditions we mentioned remain valid in both cases.

3.5 Manipulation

Manipulation or influencing the farmers' knowledge level and attitudes without the farmer being aware can be carried out if:

- we believe it is necessary and desirable for farmers to change their behaviour in a certain direction;
- we think it is unnecessary or undesirable for them to make independent decisions;
- we control the techniques to influence farmers without them being aware of it; and
- the farmers do not actively object to being influenced in this way.

In this situation the people exerting influence bear responsibility for the consequences of their actions. At times they may have their own interests in mind, as we find in many commercial advertising campaigns and in political propaganda. However, it is possible to have the best interests of the farmer in mind, as we find in many government-sponsored health and safety campaigns. Dangerous chemicals are widely used in agriculture to control plant diseases and insect pests. Most farmers would agree it is in their best interests if extension agents influence them to use these chemicals safely and correctly.

Extension also has an important role to play in making farmers aware of subtle or hidden attempts to influence them by people who stand to gain financially. For this reason the extension services in industrialized countries publish reports on official and impartial tests of tractor and farm machinery performance. Farmers then can check these performances against the claims made by the manufacturers in their advertising campaigns.

The methods discussed so far are directed at influencing farmers themselves. Important changes often can be achieved by directing influence at the farmer's situation. The next methods to be discussed are examples of changes to the farmer's situation.

3.6 Providing means

We can apply this method under the following conditions:

- the farmer is trying to achieve certain goals which we consider to be appropriate;
- the farmer does not have the means available to achieve these goals, or he or she does not wish to risk using these means; and
- we have these means and are prepared to make them available to the farmer on a temporary or permanent basis.

Specific means in agriculture include short- and long-term credit for the purchase of land or inputs such as fertilizer or production subsidies. Correct and timely application of these means, which usually are financed by public funds, may generate large increases in individual farmers' incomes. While this may help distribute wealth more widely among the population, it also may concentrate the wealth among those with the greatest power or influence in obtaining the means. Costs of providing the means may be recovered through higher taxes on the increased incomes, although there is also a danger that loans and other means will not be repaid or repaid in full, thus making them an expensive form of influence if not carefully controlled and supervised. The 'providing means' approach can be a temporary measure for stimulating farmers to try an innovation.

Government departments, including the extension service, use this method for making financial and physical means available. In several countries this has created problems for extension agents who lose farmers' confidence if they cannot distribute these means. It is more difficult to convince farmers that knowledge is also an important resource for successful farming. Even if the extension service is not directly involved in distributing credit and inputs, it has an important role to play in drawing farmers' attention to the availability of these means for improving their situation. Extension agents also can help farmers apply for subsidies, credit, etc. and assist them to make good decisions about when to use these means.

3.7 Providing service

This may involve taking over certain tasks from the farmer. The method can be used if:

- we have the knowledge and/or means available to perform the task better or more economically than the farmer;
- we agree with the farmer that it is useful to perform these tasks; and
- we are prepared to perform them for the farmer.

Income tax assessments, loan and subsidy applications, and other lengthy forms are an important but time-consuming part of modern farming. Many farmers find it difficult and tedious filling out such forms, and often are very pleased to receive free help from extension agents. However, if free assistance is given indefinitely farmers are likely to become dependent and less self-reliant. Clearly, it is in the best public interests if they can learn to complete these tasks efficiently themselves, or be prepared to pay other specialists to help. The extension agent's role in this situation is to give initial help or training in how to complete the tasks, or to direct the client to appropriate sources of professional assistance such as chartered accountants, farm consultants, etc. There are cases where farmers are considered to be incapable of learning how to perform the task themselves. Often we assume that veterinarians are the only people who can learn how to treat animal diseases, whereas farmers can learn how to treat plant diseases.

3.8 Changing the socio-economic structure

Methods for changing the socio-economic structure in rural areas may be important means of influence when:

- we agree with the farmers about their optimal behaviour;
- the farmer is not in a position to behave in this way because of barriers in the economic and/or social structure;
- we consider changes in this structure to be desirable;
- we have the freedom to work towards these changes; and
- we are in a position to do this, either through power or by conviction.

Attempts to change social structure usually will be opposed by some individuals or groups, especially when they think these changes will lead to them losing power or income. Farmers who join together in an organization may have sufficient power to overcome this type of resistance.

Extension agents can help farmers understand how economic and social structures influence their prospects for making a better living and enjoying a more comfortable style of life. They also can help them to explore ways of changing the structures or situations which prevent them from enjoying a better life. Extension agents can help farmers to predict their chances of success and to foresee possible consequences of any action they may take to change their situation, by giving them deeper insight into the social and economic forces which influence them. Extension and community development workers have helped many poor and disadvantaged people win a more equitable position in their society by showing them how to participate in political processes at local and national level. The FAO has emphasized participation of small farmers in training and development projects which help these relatively powerless people to form self-help groups for improved distribution of inputs and marketing of produce (1).

3.9 Chapter summary

We have seen that methods of influence vary according to the degree of harmony or conflict of interest between those who influence and those who are influenced, the extent to which both parties are aware of any conflict of interest, and the amount of power each possesses. It is important for the farmer and the extension agent to be aware of their common interests in an extension topic. Each depends on the other, with a change by either one possibly destroying a mutually beneficial relationship. It is usually easier for the farmer to break this relationship as he or she is not constrained by the same ethical considerations as the extension agent. However, extension agents have a potential source of power that they could misuse by virtue of their specialized knowledge. Their hold on farmers could be even stronger where they combine their advisory activities with supervision of credit, distribution of inputs, enforcement of regulations, etc.

We have pointed out how extension may be combined effectively with other methods of influence. The question is whether the same person can and should use different methods to influence farmers. Agricultural extension services in some countries are one of the few government departments to have a well developed staff and administrative infrastructure covering the whole country. Administrators sometimes find it convenient to give extension agents additional duties to their normal educational and advisory activities, including distribution of inputs, procurement of agricultural products to supply national processing and manufacturing industries, supervision of credit, and collection of statistics. It may be easier to give an extension agent an extra job than to start a new service for that job. However, the key question is whether this new job and its associated potential for influence will have a detrimental effect on the level of confidence and trust necessary for extension agents to carry out their extension activities. There is considerable evidence that these extra duties create conflicts between the role of an extension agent as an adult educator and helper, and that of supervisor, inspector or collector of debts or distributor of other types of agricultural input. In Section 10.7 we will discuss the advantages and disadvantages of giving extension agents tasks for which they have to use influence strategies other than giving advice and changing knowledge and attitudes, for example, providing means or coercion.

☞ Discussion questions

1 Imagine you live in a country in which many people suffer from an acute protein deficiency. The plant breeders at your national research institute have developed a new grain variety which has a 30 per cent higher level of protein content than other varieties commonly sown by local farmers. What are some arguments for and against using the different methods of influence to promote more extensive use of this new variety? Can you think of other methods of influence we have not discussed in this chapter? What role can extension play in each of these methods?

2 An irrigation authority wishes to motivate farmers at the upper end of the irrigation canal to use no more water than they really need for growing good crops, so that farmers at the lower end of the canal have enough water available. Which strategies can the authority use to reach this goal? Which role should extension education play in these strategies?

Guide to further reading

Smith, P. (1989) The management of change. In: *Management in Agricultural and Rural Development*, Chapter 7. Elsevier Applied Science, London.

4 Extension Ethics

'If the wrong man uses the right means, the right means work in the wrong direction'
(Chinese saying)

Extension agents' ability to influence farmers increases partly as a result of developments in communication and information technology and partly as a result of using social sciences in extension. This ability can be used not only in the interest of humanity, but also to cause serious harm, a development which implies that extension agents must take added responsibility for their actions. We see similar developments in other fields. Current military technology is much more capable of damaging humanity than the bow and arrow. Hence each extension agent is confronted with ethical issues. We cannot say how he or she should deal with these issues, but we will raise some ethical questions for which the extension agent and the extension organization should find answers which satisfy their consciences.

The behaviour of extension agents is influenced by the organization which employs them, the farmers for whom they work, their colleagues, the professional values in which they have been trained and by the wider society to which they belong. This is illustrated in Figure 4.1 by the square of

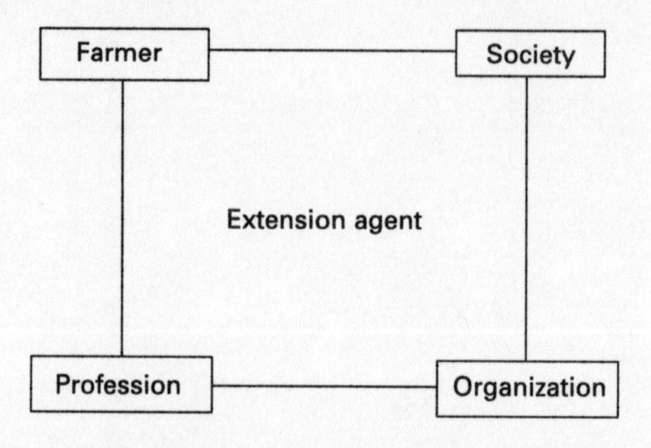

Figure 4.1 An extension agent's square of loyalties

loyalties, which is based on Proost's ideas (1). Each of these groups of people has ideas about how extension agents should behave, but these ideas may conflict with each other to some extent. The group to which the extension agents listen most will depend partly on their values and partly on the power the different groups have over them. These values determine how important it is for the extension agents to win a promotion and other material rewards, and how important to gain recognition for their work from the different groups.

We will discuss several decisions an extension agent must make, and raise several questions about which choice is desirable. We cannot answer these questions for you, the reader. You will have to answer them for yourself.

4.1 An ideal image of man and society

How do we form an ideal image of man and society? We have pointed out already how an extension agent often strives to help farmers form more considered opinions about their situation and to take more rational decisions, making optimal use of scientific research findings. In the past we assumed that the application of science automatically led to an improvement in our lives. The modern association between science and warfare has caused many people to reassess this view. Some now believe that non-scientific solutions to problems almost always are better. We do not share this extreme view. We believe that science has been used more often to improve society than to bring undesirable changes, although we accept that the latter also occur. For example, development of the insecticide dieldrin gave extension agents a valuable tool for controlling many crop and pasture pests. Unfortunately it also led to the deposition of harmful residues in the meat of animals destined for human consumption.

Continued economic growth and higher per capita incomes have been goals of most industrial societies for many years. Diminishing natural resources and increasing problems with pollution and environmental damage have stimulated many societies to reassess these goals. The catch is that many members of these societies want the advantages of economic growth without accepting or having regard for the disadvantages. Pressure groups may fight for higher incomes for their members, but show little concern for the environmental impact of increased production and consumption.

The decision whether or not to stimulate further economic growth will depend partly on the present standard of living enjoyed by the majority of members of that society. Those of us living in industrially advanced countries probably will answer this question in a different way from those living in poorer countries. Some people from industrialized countries who work in less industrialized countries assume that changes they would consider desirable in their home society also would be considered desirable in their host country.

However, even apparently fundamental questions relating to the improvement of health and prolongation of life may be answered in different ways by different societies. Our own opinions about desirable changes are influenced strongly by our personal views of a good society. These views are the result of extended cultural development processes and of personal experiences. Most extension agents have had more formal education than their farmer clients, and often have had more extensive experience of urban life and values. Hence we must expect them to have slightly different views about their societies from those held by their farmers. The differences between extension agent and farmer may be even more marked in cross-cultural situations.

The stresses and by-products of modern industrial development, coupled with growing awareness that rich countries continue to get richer while many poor countries remain poor, have stimulated many people to reassess the norms and values of their societies. It is perhaps one of the most essential tasks for an extension agent to stimulate farmers to think about these issues also. One objective of extension education and personal development training is to enlarge personal autonomy by increasing the individual's freedom to make his own decisions. At times we forget that increased freedom also brings heavier responsibilities for the consequences our actions may have for our fellow man. Furthermore, greater freedom of action also can be used to restrict other people's freedom or to harm their interests.

We can compare the ideal vision of man and society with the general norms and values of our society, and with the personal norms and values of extension agents and of their farmer clients. Unfortunately, there are likely to be some major differences between these three sets of norms and values which can lead to clashes between them. However, the extension agents who have thought about their choices will be more capable of accepting responsibility for the decisions they must take in their work.

Population control and family planning have become important issues in many countries, especially in densely populated places such as Java and Bangladesh. However, we find there are many clashes between various sets of beliefs, norms and values which make the implementation of population control extremely difficult. On the one hand, government may decide to promote family planning in rural areas because the country cannot support more people from its agricultural resources. On the other hand, the norms and religious values of the people favour large families. Furthermore, farmers place special value on having many children because of their importance as a labour resource and as a form of insurance for the parents in their old age. Government programmes to promote family planning then lead to serious clashes in personal values, as we have seen in India during the 1970s. Extension agents in these situations face some extremely difficult decisions about how to match their own values with those of their rural clients and their employers.

There is a lower probability that extension agents will try to push farmers in an undesirable direction the more they understand them, their values, their personal situations and the social structure of their society.

4.2 Loyalty

Will extension agents work and remain working for a certain extension organization? The answer to this question will depend on the extent to which they can agree with the goals of the organization and whether they think the organization is working with the correct target group. It will depend also on how much change they think they will be able to make to the organization itself and to their own way of working, as well as on the alternatives they have. It is most unlikely that any single organization will accommodate all the extension agents' ideals. The question is, which deviations from these ideals will they find acceptable and which will be unacceptable?

The extension topic also plays an important role. Extension agents may agree with their organization that it is appropriate to raise agricultural production in their districts. Such a broad goal might be achieved in a number of different ways. For example, they could be asked to work closely with a small number of large and prosperous farmers who already produce the greatest proportion of food in the area. Or they could be required to concentrate their efforts on the small farmers who previously had shown less interest in using improved methods. The extension agents' own views of the society may determine whether they are prepared to work with one method or another. The extent to which extension is trying to change the structure of the society is an important question here.

Choice of target groups is particularly important if the different groups have opposing interests. The extension agents must choose which interests they are to serve, and decide on what grounds they make this choice. The farmers themselves often make this choice because the extension agents give advice to those who ask for it. They are unlikely to give advice or information to everyone who could use it, for the simple reason that they do not have enough time. Furthermore, extension research has shown that most advice is sought by educated farmers with high social status. Therefore, if the extension agents are concerned about decreasing differences in income, knowledge and power in the society, they should be giving more attention to those who do not ask for help.

Prospective employees of an extension organization should ask carefully about the goals, policies and target groups of the organization before deciding whether they are prepared to work within the constraints of these policies and towards meeting these goals. Unfortunately, the finer details of goals and policies seldom are written down, but have to be learned by observing what the organization actually does. Furthermore, in a dynamic organization these goals and policies must be adapted continuously to meet changing needs of the society. These changes eventually may lead to conflicts with the extension agent's own goals and values, although changes to both usually are so gradual they are not noticed. We have pointed out already that extension agents are unlikely to find an extension organization with goals and policies which match their own views completely. Therefore they must estimate the extent to which they can influence the organization themselves. It is often possible to have more influence working from within an organization than by remaining outside. For example, an extension agent who can influence the adoption of small changes to the development work of a large international organization like the World Bank probably will achieve more in terms of satisfying his or her own ideals than one who works for a small voluntary organization where the goals already nearly match his or her ideals.

It is usually much easier not to promote civil servants than to discharge them. Sacking a tenured civil servant for poor job performance is almost impossible in many countries. Hence, extension agents who are prepared to forego promotion will have considerable latitude to act on their own initiative. However, the more radical their actions, the less likely it is they will influence the whole organization, because they are likely to lose the confidence of their fellow workers, and especially the directors of the organization.

The differences they are prepared to accept between their ideals and reality depend also on the alternatives available. If they do not take this job is there another one which will give more scope for achieving ideals? Have they the means to earn a living if they do not accept this job? These are particularly pertinent questions for extension agents in less industrialized countries where job opportunities often are severely limited.

4.3 The best way to help

In order to decide what is the best way to help farmers, the extension agent will have to consider similar points or when choosing an employer. The helping process is concerned with questions such as:

- which problems will they alert farmers to?
- how much anxiety should they create, and does it serve a useful purpose?
- what do they do when the results of a change they wish to achieve are only partially known?
- which criteria do they use to evaluate alternatives?

There is generally more chance of finding a solution to a problem the sooner a farmer is aware of it. Soil erosion is much easier to control in its early stages than when it has done substantial damage already. However, the extension agents have an extremely difficult task when they recognize that a farmer has a problem for which no solution is known, as is the case with some diseases.

This raises the question of how extension agents can use anxiety fruitfully in their work. People generally appear to change their behaviour faster when they are slightly, but not very, anxious, and are presented with a practical way of reducing this anxiety. For example, farmers may be told that certain sprays are very dangerous to their health (which presumably raises their level of anxiety), but that they will be safe if they use the correct mixture, wear protective clothing and wash their hands after spraying. We must always consider whether the behaviour change we seek is more important than making the farmer anxious.

Most results of each alternative course of action in an extension programme are only partly known. Should the extension agents explain these uncertainties to the farmer? What do they do if the farmer asks for concrete advice? When they explain the uncertainties, is it because it is in the farmer's best interests to understand them, or is it because the farmer might hold the extension agents responsible if results are not what had been expected? In other words, are the extension agents merely protecting themselves against criticism? Earlier we mentioned the unanticipated effects of applying dieldrin. The farmers whose pastures were contaminated decided to sue for damages the extension agents who gave the bad advice.

Are the interests and values of the farmer the only criteria when evaluating the results of alternative courses of action, or do those of the extension agent and others concerned also play a role? Farmers often will ask the extension agents what is best for them. Should the agents answer this question? If so, why? Is it perhaps because this flatters them? Do their own values influence the advantages and disadvantages they give, and the way they talk about them? Agricultural extension agents often discuss the implications an extension programme may have on farmers' incomes, but seldom those which might affect their status or personal feelings. Why is this? Is it important that the extension agent's superiors would prefer the farmers to select certain alternatives? If interests other than those of the farmers are important, what attention must be paid to these? For example, the application of phosphatic fertilizers usually will increase crop yields for farmers. However, in some areas surplus phosphate is washed into streams and lakes, causing eutrophication problems.

People from western societies working in less industrialized countries often assume, consciously or unconsciously, that their own values are better than those of the country they are working in. For example, many assume that a marriage based on affection between partners is better than a marriage arranged by the parents, without having any evidence that their assumption is correct.

4.4 Relationships

Northouse (1992) considers four dimensions to be important in the relationship between health care professionals and their patients. We list them below as they also are relevant to agricultural extension agents in their relationships with farmers.

(1) Extension agents should help farmers to make decisions which are *beneficial* to them. But on the basis of which values do they decide what is beneficial?
(2) They should not be *paternalistic*. That is, they should only provide help which farmers want. But what should they do if farmers do not realize that their present way of farming will cause serious problems for them at some time in the future?
(3) They should promote farmers' *autonomy* to enable them to decide for themselves how they wish to develop their farms. But what should they do if farmers ask them to make a difficult decision for them? Is it not paternalistic to say that the farmers should make this decision themselves?
(4) They should be *honest*, but what happens if they say they are convinced the farmers are losing money because they are not good entrepreneurs? It makes a difference whether the extension agents do not say what they think, or whether they say something to please the farmers which they believe is not true.

An extension agent certainly should consider these dimensions when deciding on his or her relationship with farmers, but difficult ethical decisions are involved in applying them. They may conflict with one another, which could have undesirable consequnces for the farmer, or for the farmer's trust in the extension agent.

Another important question relates to the circumstances under which an extension agent may pass on to other farmers information received from one of his or her clients. Who decides which information is confidential and which is not? The farmer who provided the information in the first place, or the extension agent?

4.5 Changing the structure of society

Under what circumstances and towards which goals may or should an extension agent co-operate in changing the structure of society for the benefit of farmers? The answer to this question depends on whether you are willing to change this structure through conflicts or only by giving advice to those who have to make important social decisions. Such a decision clearly would have to be made in a village development project which involved relationships between village power-holders, the extension agent and organizations of poor farmers and landless labourers.

Our view of society has considerable influence on whether we believe a conflict or a harmony of interests exists between different groups in that society. An extreme view either way in our opinion is evidence of very little insight into the reality of society. Conflicts exist side by side with harmony of interests in most societies. Therefore, it seems essential first to analyse the structure of that society.

Extension agents must consider not only the changes in society they may assist with, but also the way in which they will give this assistance. Should they join in demonstrations and other forms of social action which put pressure on politicians and administrators? What if these actions are unlawful? Should they go against the orders or wishes of their superiors, even if the superiors are unaware of the agents' action? Would we answer these questions in the same way for a development agent promoting changes in his or her own society or in another society?

4.6 Chapter summary

Extension agents may face difficult ethical choices because:

- their own values and norms differ from those of their clients;
- superiors expect them to promote extension messages which, in the opinion of clients, are not in their best interests; and
- there are conflicts of interest between different population groups.

☞ Discussion questions

1 What requirements or conditions must an extension organization satisfy if you are to work for them? Why?
2 What ethical choices do you face if you wish to give help or stimulate subsistence farmers to switch to commercial agriculture?
3 Do you consider the acceptance of non-materialistic values desirable in your country? What can/should an extension agent do to stimulate/prevent this?

Guide to further reading

Christians, C. and Traber, M. (eds.) (1997) *Communication Ethics and Universal Values.* Sage, Thousand Oaks.

Hellriegel, D., Slocum, J.W. and Woodman, R.W. (1992) Ethical decision making. In: *Organizational Behavior*. pp. 639–49. West Publishing Co., St. Paul.

Keith, K. (1995) A landcare ethic: Something of value? In *Participative Approaches for Landcare*, (eds. S. Chamala and K. Keith), Department of Primary Industry, Brisbane.

Northouse, P.G. and Northouse, L.L. (1992) *Health Communication: Strategies for Health Professionals*, 2nd edn. Appleton and Lange, Norwalk.

5 Theoretical Background to Farmers' Use of Extension

'There is nothing so practical as a good theory.' Kurt Lewin

In this chapter we will describe briefly some of the research traditions which are important for understanding how extension is used by farmers. First of all, farmers must perceive that information is available. Hence, we review some of the characteristics of perception. Then we review aspects of communication theory which give us insight into extension processes. Communication messages from extension agents often are intended to promote farmers' learning processes. Hence it is important also to understand these processes. Much research has been conducted in learning, but there is no integrated and generally accepted theory yet. For this reason we restrict our discussion to two of the most important learning theories. There is also much research on the relationships between attitude change and behavioural change. We review this research briefly and discuss its importance for extension. Decision-making in extension often is related to the adoption of innovations. There has been much research into the process of the adoption and diffusion of innovations, which has provided an important basis for extension education.

5.1 Perception

Although we live in the same world and receive similar impressions of it through our eyes and ears, and to a lesser extent through our senses of touch, taste and smell, we interpret our experiences differently.

Perception is the process by which we receive information or stimuli from our environment and transform it into psychological awareness. Extension agents cannot be expected to understand the complex psychology of human perception, but they should appreciate why people interpret their surroundings differently, and how these different perceptions influence their communication behaviour. In Chapter 6 we discuss extension methods, including the use of audio-visual media such as slides, film, field demonstrations, etc. Extension agents should be able to plan and use aids more successfully in their programmes if they understand some basic principles of perception. They also should be able to make better decisions about which alternative communication strategies to use in extension programmes.

5.1.1 General principles of perception

Relativity

Our perceptions are *relative* rather than absolute. Although we may not be able to judge the exact weight or surface area of an object we may be able to tell whether it is heavier or lighter, or larger or smaller than another similar object. When we enter a darkened room during the screening of a film we will see only the image on the screen and the bright light from the projector. After a minute or so we will be able to see other people in the room. In other words, our initial perception of darkness in the room is relative to the amount of light outside. Hence, when designing messages we should remember that a person's perception of any part of the message will depend on the segment immediately preceding it. When a designer is preparing a printed message and wishes to draw the reader's attention to a change in the message the designer leaves a blank space or changes the size of type. Perception of a message also will be influenced by its surroundings. A circle surrounded by larger circles will look smaller than a circle of the same size surrounded by smaller circles. See 'Organization' (next page).

Selectivity

Our perceptions are very *selective*. At any moment our senses are receiving a veritable flood of stimuli from the environment around us. We see objects, hear noises, smell odours and so on. Despite its capacity to process vast amounts of information, our nervous system cannot make sense of all the stimuli available. Hence an individual pays attention only to a selection of these stimuli. Several physical and psychological factors, including attitudes (Section 5.4), influence what he or she selects or pays attention to. Communication specialists who understand these factors will be more likely to draw the receiver's attention to those parts of the message they wish to emphasize. They will know which parts of the message to accentuate, repeat or diminish in order to free the receiver's mind from excess or unwanted information. We discuss selective processes in greater detail in Section 6.1.1.

Past experience also influences our selectivity of perception. Farmers who have worked with livestock for many years will be much more aware of small differences in body shape, quality of wool or fur and general condition of the animal than a person who is unaccustomed to working with livestock. Training is a way of providing an organized and structured set of experiences to influence our perceptions. For example, the agricultural student who has received training in agronomy and botany will see a pasture as a collection of specific plants, some of which have high nutritive value and others of which might be useless weeds. The untrained observer, on the other hand, may see it simply as a patch of grass. The important point for the extension specialist is that training changes your perceptions from those of most other receivers, usually in the direction of greater sensitivity to detail. Thus training may lead us to have different perceptions of situations from those of untrained people. This may create problems if we assume we are 'seeing' or 'hearing' the same thing as those around us.

Organization

Our perceptions are *organized.* We tend to structure our sensory experiences in ways which make sense to us. We try to convert the 'booming, buzzing confusion' into some meaningful order. One form of organization is into figure and ground. In a fraction of a second our senses sort out visual and aural stimuli into figures which stand out from a background. A good 'figure' attracts attention, so a designer may wish to incorporate it in a specific part of a message. Our interpretation of the 'figure' often will be determined by the 'ground'. Hence we might interpret a picture of a man with dirty face and hands and old clothes as a lazy or very poor person. On the other hand, we might interpret the picture as one of a hard-working farmer if it included a farm yard in the background. The moon on a cloudy and windy night provides us with an excellent example of a 'figure-ground' illusion. At first sight the moon appears to be racing through the clouds across the sky, whereas if we watch it more carefully we realize that the moon is stationary and the clouds are moving. The figure and ground effect is shown in Figure 5.1.

Another characteristic of perceptual organization is termed 'closure.' The perceiver tends to close or complete what he or she perceives to be an open or incomplete figure. Examples of visual 'closure' are shown in Figure 5.2.

Direction

We perceive what we expect or are 'set' to perceive. Our *mental sets* influence what we select and how we organize and interpret it. Set is an important perceptual concept which can be used by the communication designer to reduce the number of alternative interpretations given to a stimulus. For example, the writer of an extension bulletin who starts with a brief summary will 'set' the reader to seek the key points in it. A caption or a heading in a slide presentation 'sets' the viewer to observe those points. Similarly, a narration or description which accompanies a film or video will 'set' people to perceive the important elements selectively.

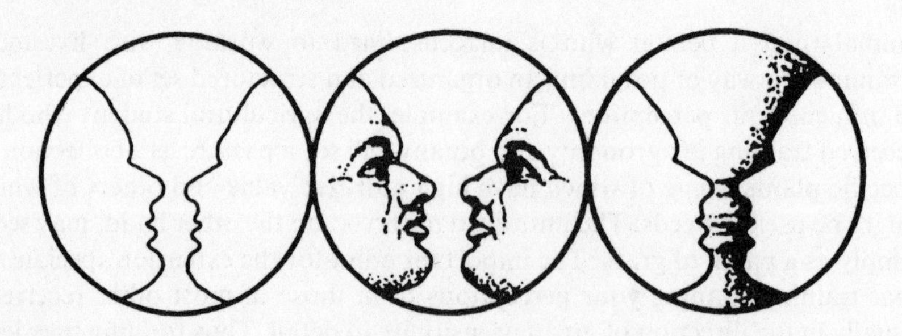

Figure 5.1 Figure and ground

There are insufficient clues in the left-hand illustration to indicate which is 'figure' and which is 'ground'. Hence we may interpret the picture as a pair of faces or a vase. On the other hand, the designer has influenced our interpretation of the middle and right-hand illustrations by adding extra detail.

From: Fleming, M. and Levie, W.H. (1978) *Instructional Message Design*, Fig. 1.8, p. 43. Educational Technology Publications, Englewood Cliffs.

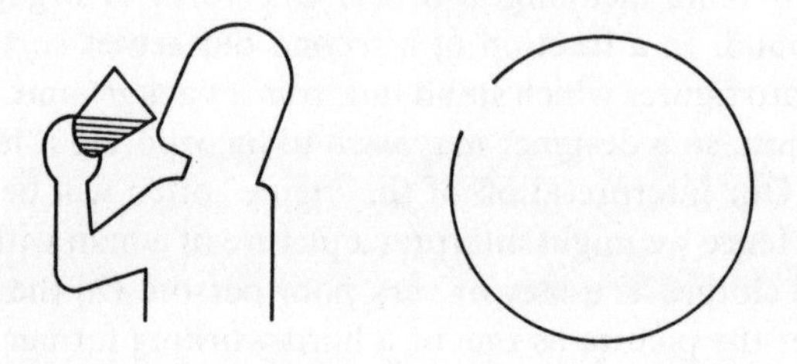

Figure 5.2 Closure

We tend to perceive incomplete shapes as closed, or we experience insight such as with the left-hand illustration showing a person drinking.

From: Ball, J and Byrnes, F.C. (eds) (1960) *Research, Principles and Practices in Visual Communication*, p. 59. National Education Association, Washington DC.

The two perceptual principles of organization and set can be combined as an *advance organizer* when preparing teaching materials. This is a piece of information which serves to explain and integrate the material which follows. It helps to sensitize the perceiver, 'setting' him or her to select the information and to interpret it in the way intended by the communicator. The act of asking a specific question in an interview may direct the respondent's attention in a particular direction. Hence, when designing a questionnaire it is important to avoid a set response by moving from general to specific questions.

Perceptual set may be a major deterrent when communicators want their audience to view or interpret a situation in a new way. We tend to respond to stimuli through habit, and these habitual responses must be broken if we are to perceive a new situation in a new way. This is a common problem with extension agents who work with people from different cultural backgrounds. Agents usually come from more highly trained and urbanized backgrounds in which they have learned to perceive agricultural conditions in a certain way. Their clients may perceive conditions differently. For example, they may place strong emphasis on mystical phenomena such as the phase of the moon when planting crops. The agent must learn to understand these perceptions before trying to change them. Experience in India has shown that rural people often have difficulties understanding Western style pictures, because these use a quite different visual code from the religious pictures they are used to in their temples. At the same time Westerners often find these religious pictures difficult to understand.

Cognitive style

One individual's perceptions will differ markedly from another's in the same situation because of different *cognitive styles.* Our individual mental processes work in distinctly different ways depending on personality factors such as our tolerance for ambiguity, degree of open- and closed-mindedness, authoritarianism, etc. Clearly it is impractical to design different messages which take into account all combinations of cognitive styles among our audiences. Hence we should adopt a strategy by which we present the same idea in a number of different ways which will appeal to most cognitive styles. This is known as message *redundancy*.

Many visual effects or optical illusions depend on the fact that we cannot directly perceive space, but must infer it from cues. These illusions normally are restricted to a class of perceptions which contain estimates and evaluations that do not correspond either with measurement or critical observation of the stimulus. For example, we tend to over-estimate the length of the vertical line in Figure 5.3, whereas in fact both vertical and horizontal lines are of equal length.

Perceptual development tends to proceed from emphasis on concrete features of objects or situations to more abstract characteristics of groupings, patterns and relationships. The communication designer therefore should

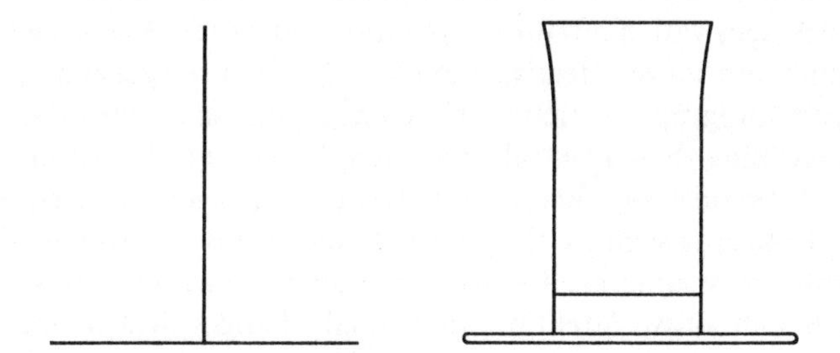

Figure 5.3 Optical illusions

choose instructional methods which are compatible with the audience's skills. Extension agents in less industrialized countries often work with illiterate farmers, and hence should concentrate on presentation of concrete rather than abstract information. In some societies farmers have had little or no exposure either to pictures or printed material. Extension agents in these situations have to be very careful with use of pictorial messages as the farmers may not interpret their meanings in the same way (1).

5.1.2 Designing effective extension messages

Gaining and maintaining attention

An extension message is useless to farmers if they do not receive it. Even if they receive it they must pay attention to it if they are to learn from it. They may receive an extension bulletin but not read it, or turn on the radio to listen to an extension broadcast but tune in to music instead. The skilled communicator therefore has to design the message to capture and maintain attention for the duration of the message.

Certain effects to gain attention are fairly obvious. A moving object among others which are stationary will attract attention. Similarly, a loud noise where there has been silence, a bright light in darkness, a coloured object amidst black and white will attract attention. The principle in these situations is use of *contrast* or of *novelty*. An extension agent who is planning to use a set of slides should consider including an occasional black and white slide among a collection of coloured slides.

Novelty is an important factor with extension messages in less industrialized countries. Many farmers have had little or no experience of films, video, slides and other modern extension aids. Their initial attention will be very great, but it would be incorrect to believe that attention will be maintained at such a high level. Audiences become accustomed to the experience and the novelty wears off.

Using pictures or words

It is common practice for extension agents in industrialized countries to prepare messages with a mixture of pictures and words. Research has shown that pictures are more effective than words for distinguishing space. For example, a photograph or picture of a plant or animal is more likely to create common meaning than a verbal or written description. However, words are best for distinguishing factors involving time such as frequency and sequence. Concepts such as plant growth, life cycles and soil erosion which involve both space and time are best communicated with pictures and words. The words can draw attention to spatial changes which are perceived visually.

We have pointed out already that a picture is not necessarily worth a thousand words in societies where people have had little or no contact with printed matter and photographs. Posters, films, videos and other visual media often are produced by technically trained people who work from a technical viewpoint rather than the viewpoint of their audience (2). Figure 5.4 reproduces a series of drawings shown to a group of young rural Brazilians. The single picture of ant control was understood by 64 per cent, while 81 per cent understood the series. Only 11.5 per cent understood the pair of pictures which portray boiling drinking water, while 32 per cent understood the series.

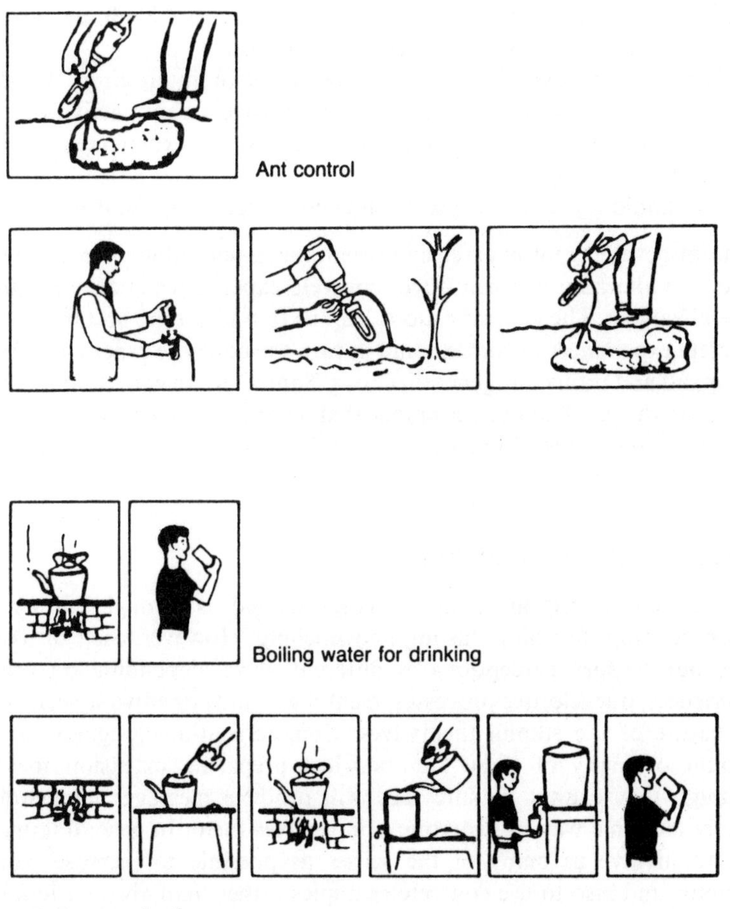

Figure 5.4 Visual information

From: Fonseca, L. and Kearl, B.E. (1960) *Comprehension of Pictorial Symbols: An Experiment in Rural Brazil*, Bulletin 30. Department of Agricultural Journalism, University of Wisconsin, Madison.

Learning by doing

We have discussed above how our past experiences influence the way we interpret information about the environment received through our senses. Our ancestors learned most of what they knew about their world through direct experiences. Farmers traditionally have learned their occupation by farming and observing the results of their labours. Most farmers in industrialized countries regard direct experience as the best, and often the only way to learn, despite their many opportunities to learn about new agricultural developments by attending short courses, reading magazines, etc.

We are more likely to develop common meanings for our experiences if they have the following characteristics:

- The senses should be strongly involved, i.e. we should all see, hear, feel, taste and smell the same objects or situations.
- The common experiences should be novel or interesting. We should participate actively rather than talking passively about activities.
- There should be an emotional component to our experiences so that we feel the importance of what we do.
- We should try to achieve something concrete, some result we can see.

Concept development in extension programmes should involve a process in which we alternate between direct concrete experiences and more abstract generalizations. The generalizations help us to understand and to place new concrete experiences in context, and in turn the concrete experiences help us to expand and refine our generalizations. Sometimes in extension we tend to move too quickly from direct practical and concrete experiences to abstract generalizations. This problem is even more acute when dealing with poorly educated farmers in less industrialized countries.

5.1.3 Section summary

We receive information about the world around us through our senses of seeing, hearing, touching, tasting and smelling. However, each of us interprets our sensory perceptions in different ways, depending on previous experiences, our selective processes, mental sets and cognitive style, as well as the nature of the stimuli themselves. Communication designers can take account of many of these factors when preparing extension messages, although they cannot be sure they will produce messages understood in exactly the same way by everyone who receives them. In general terms, it is best to involve as many of the senses as possible to increase common meaning, and also to use concrete examples rather than abstract terms.

☞ Discussion questions

1 Describe occasions where you have observed two people who perceive the same situation in a different way, for example, people who are in conflict with one another, who have a different cultural background or a different level of education.
2 Make different pictures of the same phenomenon, such as a photograph and different types of drawings (as in Figure 5.3). Ask people of different cultural or social backgrounds to describe what they see.

5.2 Elements of the communication process (3)

'It is easy to make something complicated,
But complicated to make something easy.'
Murphy's Laws

In this section we discuss first some communication models and their importance for extension. Then we look at some differences between communication channels and the relationships between the people involved in the communication process.

5.2.1 Communication models

Let us compare two situations:

(1) I have a coin in my hand and I want to give it to someone else. I do this simply by putting it in the other person's hand. Note that the coin does not change, that the other person's hand was empty before I gave him or her the coin, and that my hand is empty after I have given it to the other person.
(2) I have an idea which I wish to pass on to someone else. Is this any different from passing a coin? We think the following differences are important:

- Ideas do not become scarce. I can give away ideas without having fewer myself.
- The receiver's head was not empty before I gave him or her the idea. On the contrary, the receiver's existing ideas help him or her to appreciate my idea and to include it with his or her own.
- A coin does not change when it is passed to another person, but an idea does. An idea exists only in the human mind and cannot be transferred physically like a coin.

We can draw the following conclusions from this comparison of passing a coin or an idea: an idea must be changed into a *message* made up of several physical elements (words) with a symbolic meaning (that is to say, the idea must be *encoded* into symbols to which meaning is attached). The *source* or transmitter sends this message through a *channel* to a *receiver*. The receiver *decodes* the message (attaches meaning to the symbols) and develops an idea in his mind which he may or may not use (the *effect* of the communication). The source observes this effect and uses it to evaluate the impact of his message (*feedback*). The *words* form the main elements in the communication process and can be represented as a simple descriptive model:

This is called the SMCRE Model (Source, Message, Channel, Receiver, Effect).

Symbols and meanings

Symbols are physical elements which are important for both source and receiver. They may be verbal, such as words, or visual, such as signs, gestures or drawings. A symbol has no meaning by itself. The meaning is given to it by people who use it or see it, and, as Berlo (4) said, 'Meanings are in people'.

This generalization has an important consequence for communication. People may not attach exactly the same meaning to a symbol – source and receiver may encode and decode a message in different ways. For example, two people may use the same word but interpret its meaning differently. For most people a volunteer is someone who undertakes a task spontaneously, but for a botanist or agronomist a volunteer is a seed that germinates spontaneously. We could even share Pirandello's extreme viewpoint:

> 'All wretchedness is words. We all have a world within us, each his world with its own things. But how can we ever understand each other, sir? If I put into my words the meanings and values of that which I have in me, the other can only fill in the contents of his world! We think we understand each other, but we never do!' (5)

Pirandello considers that two people never give exactly the same meaning to words. For him, meaning and people's worlds are not the same (Figure 5.5). This is an extreme viewpoint, and we believe it is more correct to say that meanings partly coincide (Figure 5.6). The extent of overlap between the meanings held by A and B determines the extent to which they can communicate effectively, i.e. share common meanings. Hence *we can define communication as the process of sending and receiving messages through channels which establishes common meanings between a source and a receiver*.

Figure 5.5 Pirandello's opinion about overlap in meaning

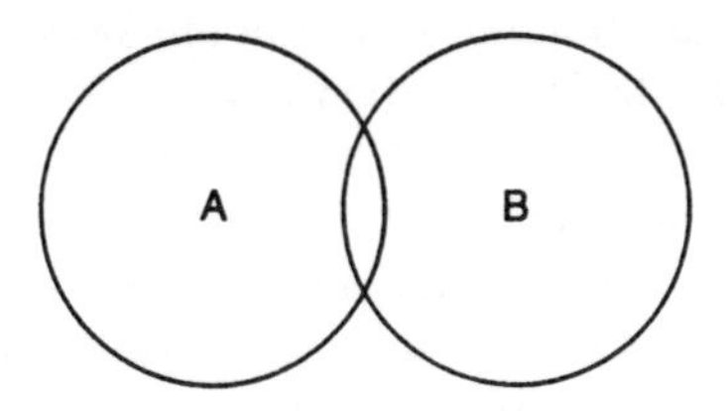

Figure 5.6 Our opinion about overlap in meaning

Getting the right message across

Receivers' expectations and attitudes influence the way in which they will decode or interpret a message. If they have learned to distrust the source it is likely that they will interpret the message in a way which confirms this distrust. Also, they may hope or expect to receive a certain message. Often they will decode the message to agree with their hope or expectations, probably more than was intended by the source. We will investigate this problem further when we discuss mass media (Section 6.1.1).

Our discussion above has some important implications for the message source:

a) Extension agents should be 'receiver oriented', that is, they should use the meanings they already share with their receivers as a starting point. Extension agents must find common ground for their discussions with farmers before they attempt to introduce new or complex ideas. This is easier for an extension agent who comes from the same rural community than for one who comes from a different cultural group. The latter will have to learn from the farmers how they express their ideas.
b) The source must use *feedback* continually to evaluate the meaning the receiver gives to the messages, and whether this meaning is the same the source had intended to arouse. This information gives the source an opportunity to change or modify the message to suit the receiver. For example, an extension agent may observe that a farmer looks bewildered or confused by an extension message. This feedback gives the extension agent an opportunity to repeat or to clarify the message.

Misunderstandings sometimes can be avoided by presenting the same ideas in several different ways ('Cognitive style', Section 5.1.1). Feedback obviously is very difficult when using the mass media. Few people will call the radio or television station to give their reaction, and those who do are an exceptional section of the audience.

A filipino medical doctor, who later became Minister of Health, has given an example of how these ideas can be used. He discussed family planning with village people by simplifying the language he used with his colleagues. However, they did not understand him. So he asked an old woman in the village how she would explain it. She said: 'The terms you use do not sound real to me, but I can understand what you are saying in terms of our agricultural practices'. Then he lived in the village for some weeks to listen to the villagers and how they talked about agriculture so that he could find analogies with family planning. There were many. So he said, for example, 'A mare that is always fed with the bran of glutinous rice will never get pregnant no matter how often it is mated. A mother who takes her pills religiously will never give birth'. Such agricultural examples only work well in areas where they are understood by the farmers. They cannot be used in a national programme, but are very effective in a regional programme if based on careful listening to villagers. Urban people gain villagers' confidence when they show that they have to learn from the villagers (6).

The SMCRE model can be very useful when analysing communication processes in extension programmes to identify the principal factors which may influence the situation. Hawkins discusses how he used a similar model to analyse the communication needs of rock lobster fishermen in Western Australia before commencing an extension campaign to control fishing pressure (7). In such an analysis we take into account:

(1) *Source variables.* The knowledge, attitudes, communication skills and social status of extension agents will influence their effectiveness as communicators.
(2) *Message variables.* The code or language of the message, as well as its content and structure, will influence effects. For example, the extension agent who designs a message for farmers about the safe use of agricultural chemicals may choose between using direct and unemotional language or emotive appeals which stress the dangers inherent in chemical sprays.
(3) *Channel variables.* We can identify whether an extension agent contacts farmers face-to-face, in groups, by print, radio, television or by a combination of channels.
(4) *Receiver variables.* In much the same way as with the source, we find that receivers' communication skills, attitudes, knowledge and social background influence how they receive and interpret a message.

Despite paying considerable attention to characteristics of the source, receiver and channel, extension research workers largely have ignored message variables. There have been many studies of communication channels. For example, we find reports with questions like 'Have you a radio?', 'How often have you listened to your farm radio programme in the last week?', 'From which channel did you first hear about the new extension idea?', and so on. We know there is a positive relationship between farmers' adoption of new farm practices and how often they read their farm magazines. However, we do not know the content of the message received by these farmers and the influence it had on their behaviour. We can ask if it is important to know what articles people read or how much contact they have with the extension service. We also have to consider the possibility that it is not so much the articles which influence the readers, but more the readers' attitudes towards certain innovations which influence the articles they read. Thus, a farmer who wishes to use new rice varieties soon after they have been released is likely to search for articles on this subject.

It follows from what we have said about the SMCRE model and from the statement 'Meanings are in people' that processes of influence through different communication channels vary from each other in three important ways:

(1) The extent to which receivers are free to interpret or decode a message according to their own views. For example, a receiver who decodes an unillustrated article about a country he or she has never visited will be more influenced by his or her own expectations than when seeing a television report which shows the life of these people;
(2) The extent to which there is feedback. There is much more opportunity for feedback in small group discussion than with a radio broadcast; and
(3) The extent to which receivers are influenced by their own group membership. Group influence is much stronger when participating in a discussion than when reading about the same topic in a magazine.

We should note several other points when choosing a communication channel:

(1) The extent to which a receiver is involved in activities associated with the message. In Section 5.3 we will show that increased activity generally improves a receiver's learning ability. Hence there are clear differences between discussion groups and a radio talk.
(2) The size of the audience. Broadcasts by national radio services may reach millions of inhabitants, whereas a lecture or talk may be heard by less than a hundred people. We will point out in Chapter 6 that large audiences do not guarantee extensive learning.
(3) The nature of the audience reached. Listeners to special science broadcasts of the national radio service are likely to be more interested in the subject matter and to be better educated than listeners to, say, a sporting programme.
(4) Who controls the pace of communication. You can read a book or

magazine at your own pace, but the sender controls the pace of messages from the electronic media. An audience with a low level of education needs more time to understand the message than a well educated audience. Therefore, a radio or TV programme which is understood by the latter category may be boring to the first category. Modern electronic equipment and technology, such as audio and video cassettes, and computer networks, give the receiver greater control over message pace.

(5) The cost per person reached effectively.

5.2.2 Relations between people

The SMCRE model probably is adequate for analysis of communication processes involved with reading a newspaper, for instance, but interpersonal communication is more complicated. You will have hardly any influence on the content of the newspaper, but interaction between people produces definite effects. This interaction is not included in the Berlo communication model. When we communicate with someone we always express something about how we see ourself, how we see the other person and how we see our relationship with this person, even if we express these messages unconsciously. A man often will communicate differently with his son than he does with his wife. This is evident not only from the words he uses, but also especially from his non-verbal communication – his gestures, posture, tone of voice, etc. An extension agent may communicate to a farmer in this way: 'I am an expert who knows how you should solve your problem', but also 'Let us work together to find a solution to your problem'. The many forms of non-verbal communication can be very important in relationships between extension agents and farmers where words alone may not express whether extension agents are sincere in their desire to help the farmers solve problems. We are aware of the words we speak, but we are only partially aware of the non-verbal signals or messages we send and receive.

Development of a personal relationship is a long process. Our reaction in a discussion is influenced not only by what and how something is said moments before, but also by what might have been said half an hour earlier. The relationship between a farmer and an extension agent may be influenced by advice the previous extension agent gave 10 years earlier. The extension agent might have spent much time and effort in solving a serious problem for the farmer.

If two people A and B quarrel, A often reacts in a way he or she feels is justified by B's 'misbehaviour'. B, on the other hand, feels his or her response was justified by A's 'misbehaviour'. Both parties do not look back the same distance in the communication process and as a result they have a different view of where the quarrel developed.

The way in which we perceive our relationship with someone has considerable influence on our reaction to their messages. If someone steps on my toe (a form of non-verbal communication) my reaction depends substantially on whether I think he or she did it intentionally or not. Whether my perception is correct or not is irrelevant to my reaction.

There also is a high probability that misperception will occur when A and B have little confidence in each other, or when they each use different keys for coding and decoding their messages because they come from different backgrounds. For example, many extension agents in some less industrialized countries have come from urban backgrounds and most have been educated in urban surroundings, whereas farmers clearly come from rural backgrounds and often have had little urban experience.

5.2.3 Section summary

We do not transfer ideas from person to person when we communicate. We elicit meanings. The source of the idea has to encode his or her thoughts in a message which can be decoded or interpreted in the same way. There is always the possibility of message distortion because sources and receivers do not always share common meanings for words, symbols and gestures. The probability of such distortion is reduced if the source is receiver-oriented, and if he or she uses feedback from the receiver to decide whether or not the message has been interpreted correctly. Different communication channels also have different effects. The interaction between sender and receiver is the important factor when analysing communication processes. Sender and receiver may change roles frequently in interpersonal communication. The messages one sends are highly influenced by the messages received. These messages may be either verbal or non-verbal, although feelings tend to be transmitted non-verbally.

☞ Discussion questions

1 Pirandello considered that 100 per cent accuracy in communication is impossible because the worlds and meanings of two people never coincide completely. How far do you agree with this viewpoint? Why?

2 How would the communication process differ between reading a book about a topic and discussing it with the author?

3 How do cultural differences within your society influence communication processes, for example, between different ethnic groups or between educated urban people and small farmers? What can an extension agent do to improve his or her communication with farmers who belong to a different cultural group from his or her own?

4 How do differences in social status influence verbal and non-verbal communication in your country?

5.3 Learning

Acquisition of knowledge and development of understanding are important aspects of receiving extension advice. They also are essential parts of the learning process. Hence it is important for us to appreciate the relevance of research findings to the psychology of learning for extension. This is especially important to help farmers adapt to a rapidly changing environment. Farmers also learn a lot from their own experience and their own experiments, from watching what other farmers do and from discussions with other actors in the Agricultural Knowledge and Information System. The extension agent has an important duty to stimulate and facilitate these learning processes which often have more impact on farmers' behaviour and on their ability to learn new ideas than knowledge which is taught by an 'expert'.

We will focus our attention in the beginning of this section on the consequences of rewards and punishments to the learning process based on research by behavioural psychologists. We will say more about the way cognitions are changed at the end of the section.

Prominent psychologists have shown great interest in the development and practical application of learning psychology. However, no generally accepted theory of learning has yet emerged. We will describe our view of the way people learn, but we have to warn the reader that some psychologists hold different views.

5.3.1 Definition

Learning is acquiring or improving the ability to perform a behaviour pattern through experience and practice. This definition implies a norm of the behavioural pattern we are aiming at. If you learn to swim, it can be to swim further, to swim faster or to improve your swimming style. We could have defined learning also as 'acquiring or improving a cognitive pattern'. However, we do not know whether the cognitive pattern of the learner has changed unless he has shown it in some kind of behaviour, which often will be only verbal. Several authors on learning would have deleted 'the ability to perform' from our definition. We do not do so, because much learning is aimed at changing cognitions which can be used at some future time to change behaviour. In our view, learning already had taken place at the moment these cognitions changed. 'Through experience and practice' has been added to distinguish the process from changes caused by biological growth.

5.3.2 The Law of Effect

The Law of Effect is a basic law of learning: 'An action which leads to a desirable outcome is likely to be repeated in similar circumstances'. We say here 'leads to' and not 'is followed by' as some authors would, because we consider the process of thinking about relations between cause and effect as essential in learning. If a group of people pray for rain and this action is followed by a rain shower, some might think that the prayers have led to the rain, whereas others may think that the rainfall was accidental. Only the first group is convinced that they should pray again for rain in the next drought. If people receive explanations about how their actions lead to desirable consequences, they will learn more rapidly from these actions. People try to behave in such a way that their actions have more desirable consequences, often called rewards, and less undesirable consequences, called punishments. The process of rewarding desirable actions is called reinforcement.

Rewards and punishments

What constitutes a reward or a punishment for a person depends on his or her cultural surroundings, on experience in the past and probably also on personality. Food usually is a reward, but perhaps not if politeness already has compelled you to eat more than you had intended. Recognition by people you respect also will be a reward. It can be particularly rewarding if we have been able to solve a problem which we could not solve in the past. An extension agent should listen to farmers to discover what they consider to be rewards and punishments.

Punishment has the opposite effect to reward, but is much less effective. Teaching in school and leadership in organizations often are based mainly on punishment of wrong behaviour. A system which is based on rewards for correct behaviour in the first place will be more effective.

People must be able to distinguish which actions are rewarded and which are not if they are to learn from their actions. This is often difficult in the real world because these rewards depend on many different factors which are hard to disentangle. High crop yields are a reward for most farmers, but why did a particular farmer get a higher yield this year than last year? Was it because of better land preparation, application of more fertilizers, choice of variety, more careful weeding or rain at the right time?

5.3.3 Self-efficacy

People will learn more if they try a wider range of different actions. The extent to which they will do this depends on the rewards they received in the past for trying something new. If this action was rewarded, they will try it more frequently, and they might also try actions which deviate more from the traditional way. If it failed, they will be hesitant to try something new again. Poor farmers with little education often are rather apathetic, because all attempts to improve their situation have failed in the past through lack of resources, lack of power and/or lack of knowledge of innovations. If you wish to change this attitude you should start with small changes which are successful. Any attempt to help them which fails will confirm their conviction that it is impossible for them to improve their situation and that nobody is really interested in helping them.

We speak here of *self-efficacy*, the perception people have of their ability to perform a certain task well. If people with a high level of self-efficacy or perceived behavioural control fail to obtain the desired results, they will try again and/or try to discover what they can do better. People with a low level of self-efficacy will soon stop trying. Self-efficacy is specific for certain tasks. It is important that people learn to make realistic estimates of which tasks they can or cannot perform. They will learn this partly from their experience of performing the same or similar tasks in the past and their interpretation of this experience. 'Did I fail because of bad luck or because of lack of skill?' They learn it also by observing others, who are similar to them, performing the tasks. 'If my neighbour is quite capable of calibrating his sprayer, I should also be able to learn how to do this.'

5.3.4 Learning processes

We may learn how to behave more effectively for different reasons:

- a change in our environment which leads to a different pattern of rewards and punishments for our actions;
- by observing which rewards and punishments other people get for their actions;
- information given to us about the consequences we can expect from different actions; and
- by thinking about the consequences we might expect from certain actions.

Programmed instruction is a well-known application of this research on learning processes. The subject matter to be taught is divided into many small parts with a question prepared for each part. Students often will find the answer to this question on the following line or on the next page. They receive immediate reward by progressing to the next step if they have given the correct answer. Programmes are designed so that the pupils are seldom punished if they do not find the correct answer. Many research studies have shown that good results can be achieved quickly with this method, but it requires the teacher to know exactly what is the correct behaviour. We use this method when we learn another language through a language laboratory, and we can use it also with much technical knowledge about agriculture. It is not appropriate when the goal is for the learner to discover his or her own unique solution to a problem, for example, the farming system which suits his or her goals and resources best.

Extension agents can use these ideas by improving the feedback system to help farmers learn more rapidly from their experiences. Many farmers, especially subsistence farmers, have only a vague idea about their yields. There has to be a rather large increase or decrease in yield before they notice the difference. It is important to teach them how to measure these yields more accurately. Milk testing by dairy herd improvement organizations is a well known example. The farmer also can draw better conclusions if he or she keeps accurate records of how he or she has treated this field or that animal.

Modern computer technology improves the possibility for rapid and accurate feedback. For example, a cow's milk yield depends on her age and the number of months since last calving. This makes it difficult for a farmer to say whether a cow yielding 15 litres a day is producing well or not. However, there is a computer program which standardizes production levels of all cows to that of a six year old cow two months in lactation. This enables the farmer to see whether a change to another feed increases or decreases milk production. Similar programs have been developed to help farmers to draw the right conclusions from their accounts.

Not only do we learn what consequences we can expect from our own actions, but also we learn from observing other people's actions. This is often the major way we learn from experience. Such learning will be more effective if we discuss our experiences with a person who has tried something new and interesting, then compare these with our own observations. Experiences often are discussed at length by farmers when they meet each other. The effectiveness of their learning depends on how reliable their information is about actions and their consequences. You must also translate the experiences of the person you observe to the consequences you could expect yourself with different resources and skills. Learning by observing the experiences of other farmers is important because this is a much better way of making a decision than by processing all the available information yourself. One farmer can watch carefully to see what happens to another who tries an innovation and this can be a conscious learning process, for example. It is also possible to learn new behaviour patterns without realizing it. Television programmes can have a significant impact on learning new interpersonal relations when they are watched for recreation rather than for learning.

5.3.5 The cognitive map

Much extension teaching is done during farm visits, at meetings, through articles in farm magazines, etc., by telling farmers how they can change their actions to obtain more rewards and fewer punishments. The effectiveness of the communication process determines the extent to which these activities lead to farmers learning. This effectiveness can be enhanced by structuring the new information clearly and in such a way that the farmers can integrate it into their present pattern of thinking. This pattern of thinking often is based on concrete experience and not on the abstract ideas taught in agricultural schools. Bruner (1966) and his co-workers have studied how people can be helped to change their view of how the world in which they live operates – their *cognitive map*. This map is constructed by observing objects and events, analysing the observations we make and then making generalizations from this analysis. The process of constructing a cognitive map basically is the same as the scientific research process. Bruner and colleagues assume that it is not so important at schools to teach children many facts in this time of rapid technological change. It is more useful to learn how to discover relations between facts and phenomena. They try to structure what is already known and new information in such a way that the children suddenly gain new insights after they have wrestled with the problem for some time. This breakthrough often does not take place at the moment of problem-solving, but some time afterwards when not working on any problem. However, it happens only after thinking about all the relevant information when trying to solve the problem. It is a way to change a person's cognitive map of reality.

So on the one hand learning can be influenced by changing the pattern of rewards and punishments. In rather simple cases this will result automatically in a change in cognitions of the relations between actions and their consequences. On the other hand these cognitions can change through an active process of thinking. This process is stimulated by the learner's motivation to solve a certain problem, by a clear structure of the information needed to solve it and by several attempts to apply this information to finding a solution. In this way we can teach farmers how to solve their problems themselves rather than teach them solutions for their problems. We consider this an important task for extension agents. Therefore we devote considerable attention in Chapter 6 to group discussion and counselling techniques which apply this approach to learning.

5.3.6 The motivation to learn

These learning processes require farmers who are motivated to learn, but at the same time the processes will increase their motivation. It is always more rewarding to find your own solution to a problem than to be given the solution by someone else. Motivated people tend to be very active learners, although high levels of motivation may lead to learning blocks if the learners do not succeed in their learning tasks. It is more effective if the extension agents create a situation in which the farmers themselves discover there is a need to change, rather than having the extension agents tell them. Kenyan extension agents helped a group of farmers to calculate themselves how much more they would earn from growing hybrid maize than from growing the local maize. The group decided which data should be included in the calculation, such as the yields which could be expected. Their calculation showed a big advantage of the hybrid maize. The farmers' reaction was: 'Damn! Why didn't you tell us this before?' In fact the extension agents had told them for several years already, but the farmers had not paid much attention. They worked hard for two days to learn the best way to grow hybrid maize (8). Extension agents should try to teach a subject at the moment farmers realize they will need this information. They are then motivated to learn, and are likely to remember more than if it had been taught a few months earlier. It is easier to apply this principle in crop production than in animal husbandry.

5.3.7 Effective methods of learning and teaching

Learning also takes place without assistance from teachers and extension agents. Most of our learning occurs when analysing our own actions or the experiences of others. Farmers who have seen that neighbours can afford improvements to their houses after starting to grow vegetables might be interested to learn how to grow vegetables themselves. They can observe carefully what their neighbours are doing, talk with them, talk with other people who know something about vegetable growing, listen carefully when it is discussed on the farm radio programme, etc. The extension agent can stimulate and complement these learning processes by organizing a practical course on vegetable growing in this village, for example. With an effective course he or she will build on the experience these farmers already have in growing crops and vegetables for their own use. The farmers will learn more when their experiences in their own fields are used as a learning situation rather than when they merely listen passively to the instructor. They also should be rewarded for asking questions when something is not clear, because the instructor needs this feedback to be able to teach successfully at the farmers' level of understanding. Unfortunately, in some countries people have learned not to ask a 'wise' teacher any questions, and certainly not to disagree publicly with their teacher.

The extension agent should state clearly what the learning objectives are at the outset. This makes it easier for the farmer to organize the knowledge which is presented. Then the extension agent should move step by step, beginning with farmers' current knowledge and abilities regarding the learning objective. Often it will be useful not only to present the new information and the new skills verbally, but also to demonstrate them and to give the farmers an opportunity to practise them. People with a low level of education learn better from their own observations than from reading or listening. The extension agent should not expect his farmers to deviate much from the way of learning they are used to.

When teaching farmers about fertilizers the extension agents first can demonstrate their correct use, then discuss what has been demonstrated, why it has been done this way and what the results are. They then can ask the farmers to try it themselves on a small part of their fields. Finally, they can help them to analyse the results they obtained and compare these with traditional practices.

The situation, and particularly the farmers' and the extension agents' skills, will determine how far the extension agents can go with this teaching process. For effective learning they should try to prevent the farmer from making serious mistakes such as application of the wrong pesticide.

The introduction of Integrated Pest Management (IPM) in rice in Indonesia is an interesting example (9). In Farmer Field Schools farmers learned from their own observations which insects damage their crop, which other insects are predators of these pests and how they can influence the pest population in their field. Whenever possible the facilitator does not answer farmers' questions, but helps them to learn from the processes they can observe in the field. These schools have been so successful in reducing costs of pesticides, increasing crop yields and making agriculture more sustainable, that the Indonesian Ministry of Agriculture now has adopted this extension approach for many other problems. It can only be achieved through retraining the whole extension staff who had been used to a top-down approach.

Translation of theory into action often is difficult in learning processes. Therefore presentations of new ideas by extension agents should be combined, whenever possible, with experiments by farmers, visits to farmers who have tried these ideas and discussions among farmers to exchange their experiences regarding application of the ideas.

Extension agents should not only provide farmers with solutions for their problems, but also increase their knowledge and understanding in order to increase their skills in developing solutions which work well under their circumstances and help them to reach their own goals. Educating farmers is often a more important task of the extension service than transfer of technologies. The Indonesian Farmer Field Schools are a good example of this approach. Competent farmers are a major key to successful agricultural development.

5.3.8 Section summary

Learning is acquiring or improving the ability to perform a behaviour pattern through experience and practice. People try to behave in such a way that their actions produce more desirable outcomes (rewards) and fewer undesirable outcomes (punishments) than they experienced in the past. Therefore, they have to know which actions lead to which outcomes. This requires a distinction between actions which affect outcomes in a different way, and correct feedback on the outcomes obtained. It also requires an interpretation of the relation between action and outcome.

Farmers learn from their own actions, from observing others' actions and by discussing the relationship between cause and effect. These learning processes depend on the active involvement of the farmer. The extension agents can try to create a situation which makes learning easier for the farmer. They can help farmers to become aware of what they have to learn to solve their problems, and to structure available information in such a way that they gain new insights more easily.

☞ Discussion questions

1 Extension agents who base their extension strategy on principles of learning psychology restrict the freedom of farmers to decide their own behaviour. Do you agree with this statement? Why?
2 What possibilities do you see in extension for making use of the learning theories?
3 To what extent does this textbook use different applications of learning theories:
- (a) rewarding the ability to perform new behaviours;
- (b) structuring information in such a way that new insights are developed?

5.4 Attitudes

5.4.1 Definition

An attitude can be defined as the more or less permanent feelings, thoughts and predispositions a person has about certain aspects of his environment. Components are knowledge, feelings and inclinations to act. Or to say it more simply: an attitude is an evaluative disposition towards some object or subject which has consequences for how a person will act vis-à-vis the attitude object. The emphasis now in much research is on the feelings or emotions.

5.4.2 The study of attitudes

In the past it was thought that an attitude influences a broad range of behaviours, for example, a positive attitude towards modern agriculture will stimulate the adoption of many different innovations. Therefore it was considered to be an important task for extension agents to change negative attitudes. However, in the 1960s researchers showed that the relationship between attitude and behaviour is often weak. Consequently there is hardly any relationship between change in an attitude and change in behaviour. If empirical facts do not confirm a theory, researchers have a good reason to do more research in order to develop a better theory. This was the case with attitudes. Many social psychologists now assume that, among other factors, behaviour is influenced by behavioural intentions. These intentions are influenced not only by the attitudes of people, but also by the expectations regarding their behaviour from their social environment, the subjective norms, and by their perceived ability to carry out this behaviour, the self-efficacy.

This new way of studying attitudes was developed by Ajzen and Fishbein (10) in their model of reasoned action which is reproduced in Figure 5.7. It does not describe properly spontaneous actions to which little thought has been given, and routine decisions which we mention in the next Section 5.5.

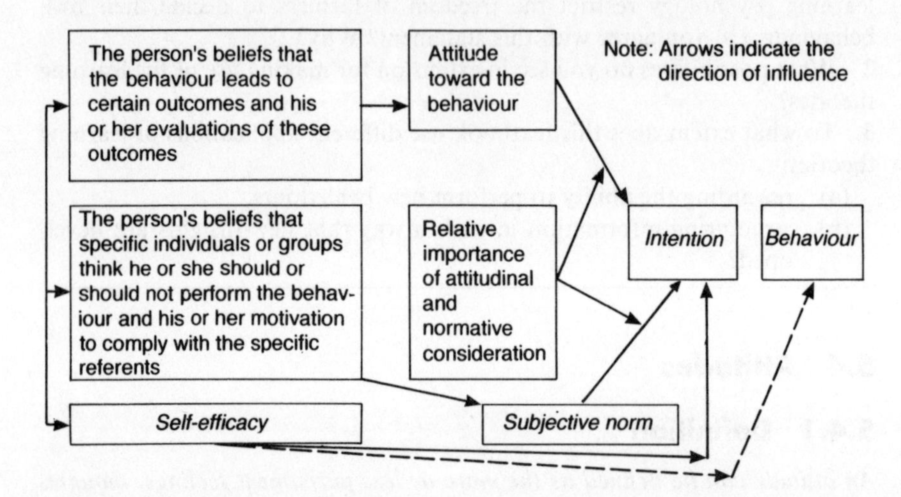

Figure 5.7 The behavioural intention model based on Ajzen and Fishbein (1980) and Ajzen (1988)

If we study attitudes in order to predict behaviour, both have to be measured at the same level of specificity. For instance, a general attitude towards modern varieties of maize will only have a weak relationship with the choice of maize variety grown by the second wife of a Tanzanian farmer to feed her children and her husband in a year that rain started late. To predict this behaviour we have to measure the attitudes towards the same behaviour and take into account that a change in behaviour usually has several favourable and unfavourable consequences. The most important consequences should be measured in order to predict behaviour. For instance, when compared with the local maize, hybrid maize may differ in average yield, seed cost, drought resistance and resistance to stemborers and streak, taste, cooking quality, seed colour, etc. One would ask the farmers which characteristics are most important to them in deciding which variety to grow and what they think about each of these characteristics. These opinions then can be summed by attaching a weight to each characteristic according to the importance it has for the farmers. It is also possible that a farmer will not accept a variety which does not meet at least a minimum standard on a characteristic he or she requires, such as very good drought resistance, for example. Most Tanzanian farmers will not accept a yellow maize, irrespective of how good it is in all other characteristics, because maize porridge should be white. This approach to measuring attitudes has the advantage that it gives clear indications of how effective extension messages can be formulated.

A difficulty with this way of measuring attitudes is that people have different goals at the same time. For example, a farmer's goals may be high income, low risk, to be well liked by his or her family, to have high status in the community and a reasonable amount of work. It can be very difficult to estimate how each characteristic of an attitude contributes to the optimization of the aggregate of these goals. (See Section 5.5.)

New attitudes which are based on the respondent's experience, or on systematic thinking about the problem, are more likely to bring about change in behaviour than more superficial attitudes. Farmers' attitudes are more likely to correspond with their behaviour if they have experimented with an innovation themselves or have collected information about it from sources they consider reliable than if they have only heard an interesting talk about it on the radio.

Our behaviour is influenced not only by our private opinions, but also by the expectations we perceive our social environment has of our behaviour. For example, one of the authors nowadays dresses more casually on Sunday than on weekdays, whereas 40 years ago he dressed more formally. An important cause of this change in behaviour is that the expectations of his friends regarding his way of dressing have changed. For certain kinds of behaviours you may be highly influenced by your parents' expectations and for others by those of your friends. These expectations are called *subjective norms*. For the measurement of these norms we have to ask a person whose expectations regarding a certain kind of behaviour are important to him or her, what kind of expectations he or she thinks they hold regarding that behaviour. The choice of a maize variety by a rural woman may be highly influenced by her perception of her husband's taste preferences. We will re-examine the importance of these norms in Section 5.6.4.

We are not always able to fulfil our *intentions*. We may consider a Mercedes Benz the ideal car, but not buy one because we are only able to afford a bicycle. Therefore, nowadays we try to take the actual situation of the respondent into account to predict behaviour from the measurement of intentions. Ajzen and Fishbein later enlarged their model by adding 'self-efficacy', a concept we mentioned earlier in discussing learning processes (Section 5.3). Self-efficacy independently influences the intention of attitudes and social norms, but it also has a direct influence on behaviour.

The Ajzen and Fishbein model provides better predictions of behaviour than previous attitude models. However, there are still some difficulties. It is not always possible to obtain very accurate information in interviews about the way in which feelings influence attitude and behaviour. For example, it is possible that farmers buy a tractor, not because they expect it will increase their income, but because it will enhance their status in the community. They may not be willing to admit this in an interview. Neither may they be well aware of these feelings themselves. Behavioural decisions often are based on only a few of the beliefs and feelings which the actor considers most relevant in the present context. Furthermore, the model assumes a very deliberate decision-making process.

Measuring attitudes and behavioural intentions in this way is time consuming and the results may only be valid for a limited time. In agriculture it will often be location-specific. The resources may not be available to base extension messages on attitude research for all topics and for all different target groups included in an extension programme.

It is not only the case that attitudes influence behaviour, but also that behaviour influences attitudes. For example, consumers may use a new product which they saw by chance in a shop or which was offered to them by a friend. The result of this experiment will influence their attitudes towards this product more than advertisements they see.

5.4.3 Model of behavioural change

This analysis of attitudes has produced a model of behavioural change (see Figure 5.8) which can be used by extension agents who try to persuade their farmers to behave in a certain way, such as to give the 'right' dose of nitrogen to their rice crop before transplanting (11). It is not a good model for extension agents who try to help farmers make their own decisions:

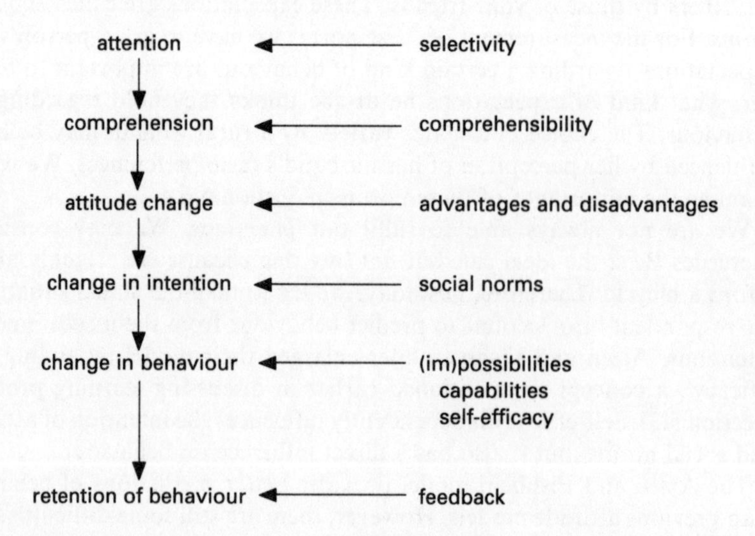

Figure 5.8 Model of change in behaviour through change in attitude

Extension agents first will have to arouse their clients' attention by presenting their message in such a way that the clients select it from the vast array of messages available to them. They may do this by including information for which the clients already feel a need. Other ways are discussed in Section 5.1.2. The message has to be presented so that clients understand it and can incorporate it into their way of thinking. The clients then should be helped to balance the advantages and disadvantages of the new behaviour. Extension agents often believe it is most effective to talk only about the advantages and to keep silent about the disadvantages. However, many clients will discover these disadvantages anyway, either from their own experience or from other sources in their knowledge and information system. They may lose confidence in their extension agent if he or she does not mention the disadvantages. Some extension agents and their bosses have difficulty in accepting that farmers are capable of deciding for themselves whether the advantages outweigh the disadvantages for them. If the extension agent comes with a message which deviates widely from his or her clients' present opinion, this message will be rejected, especially when the extension agent has not yet gained much confidence among his or her clients.

If the change advocated by the extension agent is supported by people in the clients' social environment, there is an increased probability that the clients will change their intentions. This is one reason why we discuss the role of group discussions in Section 6.2.3. It is quite possible there will only be changes in behaviour if there are also other changes in the environment and/or the farmers' skills. Perhaps fertilizers have to be made available in the villages or the farmers have to learn to spread them evenly.

It is clear from our discussion of learning processes that farmers may revert to their old behaviour unless feedback teaches them how their changed behaviour will help them to achieve their goals more effectively. For example, how it pays to use the recommended amount of nitrogen.

5.4.4 Section summary

Different models have been developed to study attitudes and the way knowledge about attitudes of the target group can be used to change their behaviour. We discussed Ajzen and Fishbein's model which is often used in education campaigns in which the educators know the direction in which behaviour 'should' be changed. It is a model for deliberate decision-making, but not all our behaviour is based on these types of decisions. It can also be costly and difficult to collect the right kind of data for this model.

☞ Discussion questions

1 What do you consider are the advantages and disadvantages of using Ajzen and Fishbein's attitude model for an extension programme on:
(a) the introduction of improved varieties;
(b) changing from the production of food crops to vegetable production?
2 Which variables would you have to measure for each of the above examples?

5.5 Decision-making

Many extension agents see their role as one of improving the quality of farmers' decision-making in order to help farmers achieve their goals more satisfactorily. Others strive to get farmers to make decisions which they consider important to help the farmers to improve their way of life and/or to achieve national goals. In both cases the extension agents need to understand the decision-making process clearly if they are to achieve their objectives. In this section we will try to develop this understanding and to stimulate thought about the role extension agents can play in improving farmers' decision-making processes. We will give a general overview of different decision-making theories and will suggest why decision-making processes do not always develop sufficiently to achieve the decision-makers' goals.

We discuss here how individuals make decisions, but not how others influence them or by whom they are influenced. Also, we give little attention to how groups make decisions. These also are important topics to which some attention will be given in other parts of this book, but which are discussed more thoroughly in textbooks on social psychology.

5.5.1 The normative model

Various research workers have tried to define what course a decision-making process should follow to achieve 'best' results. 'Best' implies the goals of the decision-maker are achieved as closely as possible. This *normative* approach assumes a sound decision-making process should pass through the following stages:

(1) Become aware of the problem. This may be because:

- the present situation is unsatisfactory, as with a disease for example;
- the decision-maker considers that continuation of current developments will lead to difficulties. For example, small farmers who can barely make a living now realize that, with the increasing cost of many things they buy, they will not be able to do so 10 years from now unless they can increase their farm size;

- the decision-maker becomes aware of new solutions to his or her problem. For example, the farmer learns that other farmers have increased yields by using modern technology.

(2) Establish what the goals are. The problem is that people often set different goals which cannot all be achieved at the same time. Hence we must set criteria against which we can match the various goals. For example, our hypothetical farmers may want a higher income and to work shorter hours. They may be able to increase their income by leasing more land but they also will increase their work load.
(3) Diagnose the causes of the problem. It is almost impossible to find a satisfactory solution to a problem if we do not understand what causes it.
(4) Review possible alternative solutions to the problem, and consider the results we might expect from each of these alternatives.
(5) Evaluate the expected results against the criteria for a good solution set out under Point 2 above.
(6) Choose the best possible solution.
(7) Implement this choice.
(8) Evaluate if we have really achieved the expected results, and therefore whether the problem has been solved satisfactorily.

Different goals

In order to make a decision we have to know which alternatives may be used and what consequences we can expect from following each of these alternatives. An extension agent can help a farmer considerably at this stage, because in theory much of this information can be obtained from research. On the other hand, we have to know what our goals are or which values we use in choosing between alternatives. The extension agents' values often will be different from those of the farmers; after all, they chose to become civil servants and not entrepreneurs.

The distinction between knowing what are the consequences of different alternatives and values is not as clear as it might appear. In the first place, what we assume to be true depends partly on our values, as we have pointed out in Section 5.1 on perception. In the second place, we have different levels of goals. A farmer might have a goal to own cows with a higher genetic potential because they give more milk, and this in turn will increase his or her income which will make him or her happier. The ultimate goal, happiness, depends indeed on values, but it is possible to predict whether the intermediate goals will be achieved by reaching the lower level goal. For example, whether higher milk production gives a higher income.

Extension agents often assume that farmers have the same goals as they have. This will usually be correct for lower level goals, for example, regarding advice on plant protection. But it is a dangerous assumption for higher level goals.

Socio-psychological values

The many variables to be considered when making a choice are a problem, apart from the difficulty of considering them all in depth. Furthermore, it is hard to set criteria for a decision when we must take into account its socio-psychological value as well as its practical outcome. For example, small farmers in a less industrialized country who want to increase their income may consider a wide range of alternatives including development of a cottage industry such as weaving, use of more fertilizer on their crops, planting inter-row crops, leasing more land or encouraging their wives to take a job. The latter choice may be unacceptable because of local customs and beliefs, even though it may yield the highest income for the family. Socio-psychological values such as these are very difficult to measure and to weigh up against income.

We assume that better decisions are made when we understand how socio-psychological values influence our choice. Status often plays an important role when we buy an expensive new car. There is perhaps less chance the buyer will spend more money than he or she can afford if he or she is aware of this. On the other hand, the pleasure gained from feeling higher in status may decrease if the buyer realizes that this was the reason for buying the car.

5.5.2 The empirical model

Empirical research into how decisions are made in practice shows that people do not always aim to maximize results. This is because we must also consider the costs of making the decision when determining the maximum result. A farmer who spends 20 hours deciding whether to use Brand A or Brand B weedkiller on the house vegetable garden instead of spending an equivalent time weeding the plot by hand may find the costs of the decision high when compared with the results. Also, we often have to make our decisions on the basis of limited information about their consequences.

When you chose your school career you probably had limited information about your future relationships with fellow students and teachers, whether you would succeed with your studies, and what job opportunities you would have. Of necessity your choice was subjective and had to be made in a changing situation. Your goals when you selected your study area were likely to be different from those you have now. Employment possibilities also may now be different. We can only hope to make a decision which can be justified at that moment, knowing our choice may not be optimal at a later time.

We often accept a decision which satisfies certain minimal demands. We do not always consider all the alternatives open to us when deciding whom to marry. An attractive person with a pleasant personality from a good family who we know well may be accepted without seriously considering the millions of other possible candidates.

It is very difficult to consider all possible alternatives and to predict the consequences of many decisions. We tend to choose the same solution we have chosen previously with routine decisions. For example, we usually buy bread from the same shop. In principle, by making routine decisions we save time which can be devoted to more innovative decisions. The catch is that we often continue to make a routine decision by habit, even though the situation has changed. We continue to buy bread from the same shop despite the fact that a neighbouring shop has started to sell bread of the same quality at a lower price.

Time and effort often make it impractical to consider all alternatives and, at the same time, predict the consequences of innovative decisions. It may be feasible to deviate only slightly from an established routine. When the director of an extension service draws up the departmental budget for the coming year he or she is likely to base it on the current budget, with only minor modification to some items. Changing all items each year is difficult and time-consuming. Furthermore, it creates confusion and uncertainty among staff in the department unless these changes are discussed with them in advance.

Farmers often have to make decisions based on probability estimates, for example, that it will rain next week, prices will go up or variety A will give a higher yield than variety B. However, it is difficult for human beings to make probability estimates correctly. The estimate of the amount of rainfall next year may be influenced more by the amount that fell last year than by the average of the past 20 years. Extension agents can play a useful role in helping farmers make their decisions on the basis of more correct probability estimates.

Decisions seldom are made on their own; they usually form part of a longer sequence of decisions. A decision we take now may influence our freedom of choice with future decisions, although we may not be able to see this point clearly at the time. Young people entering agriculture may decide to raise livestock rather than to grow crops. They then will have to decide what breeds of livestock to have, what pastures they will feed them on and how they will manage these pastures. Their choice of pasture species when establishing new pastures, and especially their choice of legumes, may influence future fertilizer requirements as well as the condition of the soil in 20 years time.

Farmers have to make this sequence of decisions in an environment of uncertainty. At the time they sow their crops they will not know how its growth and development will be affected by weather conditions, plant diseases, etc. Nor will they know at what price they will be able to sell the crop. This uncertainty will decrease as the cropping season progresses. Hence it is a wise strategy to make each decision as late as possible. Farmers should not decide in advance which set of agronomic practices to use, as is often advised by research workers. It is best to base their decisions about each practice on all the information they can gather at that moment. For example, if the crop is infested with weeds they should not apply much fertilizer because it is more likely to stimulate growth of these weeds rather than the crop.

5.5.3 Bos' model

There are reasons to doubt whether the phases of the decision-making process can be worked out in the way that people with a normative view of the process suggest. Bos has developed a completely different model of group opinion formation which appears to have relevance for individual decision-making (Figure 5.9) (12).

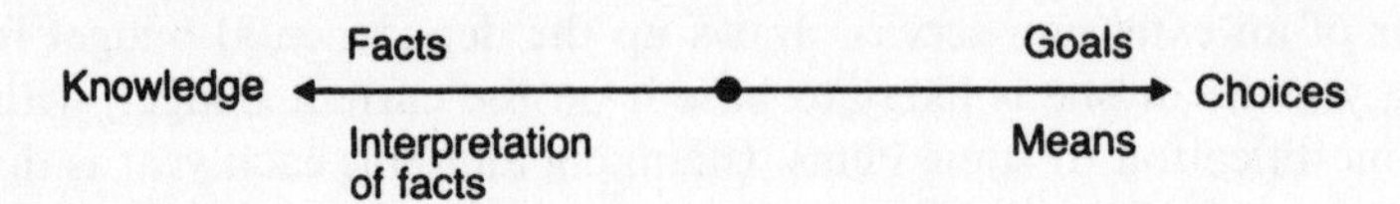

Figure 5.9 Bos' decision-making model

He points out that on the one hand we are dealing with the question 'What must I choose?', and on the other hand with the question 'What is the present situation?'. We must examine the pathways of choice where we are dealing with goals and means to answer the first question. To answer the second question we must examine the pathways of knowledge in which facts and interpretation of these facts play a role. This also is called the search process. The difficulty in making a choice is that we are uncertain what the outcome will be. In the decision-making process we use information (knowledge) to reduce this uncertainty. Choice and knowledge must be examined alternately. A general review of available choices will give insight into the knowledge we need for solving the problem. With this knowledge we can formulate the problem more accurately and can thus advance a step further towards making a choice. We cease the process at the moment of further reduction when our uncertainty becomes less important for us than the efforts we have to make to reduce this uncertainty.

During limited empirical testing of his model, Bos found it was necessary to pay adequate attention to his four fields (goals, means, facts and interpretation of these facts) in order to achieve sound decision-making. However, he found the sequence of the fields is interchangeable.

Farmers will first search their own memories for knowledge. If they are not satisfied they will search elsewhere by discussing the issue with their colleagues or with extension agents, by reading farm papers, etc. They will only spend much energy in searching for information for important and risky decisions and only when they feel that they are able to process the information they acquire properly. Proper processing will not be possible, for example, when they are not clear about the criteria they should use for choosing the right decision.

5.5.4 Type of decision-making

What type of decision-making should an extension agent strive for with farmers? We think it is possible to integrate the three approaches of the normative model, the empirical model with its limited insight, and Bos' model. The phases hypothesized by people thinking about normative decision-making should be seen as functions that must be fulfilled, but not necessarily in the order stated. Bos' model seems to be a more realistic method of fulfilling steps 2 to 6 of the normative decision-making process. However, we must remember that human information processing capacities are limited.

Problem-solving decisions require:

(1) a clear analysis of the nature of the problem and of its causes;
(2) possible solutions to the problem;
(3) a control system which shows whether the problem is solved by applying these solutions.

An extension agent can help with each aspect of the decision-making process.

5.5.5 Analysis of the problem

Timely awareness of the problem is necessary. For example, farmers might need help at the appropriate moment to be made aware of the precise nature of their problems. They may have to choose between reducing their risks and increasing labour productivity by specializing in a few crops. If they delay their decision too long they may not have the resources necessary to make a success of either alternative. Once the problem has been defined clearly, farmers usually continue the decision process without further help. This means that problem definition focuses on what they can do to solve the problem rather than on what others should do. Psychologically, we are inclined to place the blame for our problems on other people.

Gradual changes in a situation also are noticed much less than acute or sudden changes. Many people would consider a temperature of 38.5°C due to influenza to be a problem, but would not regard being 20 kg overweight as very important, despite the fact that the latter condition may have many more consequences for their long-term health.

From our definition of the concept 'problem' (see Section 2.1.3) it appears that awareness of a problem depends not only on observing the present situation but also on the desired situation. However, often we are only vaguely conscious what this goal is. Thus it is important to be able to specify the problem clearly before starting to make decisions. Extension agents can help with this task in several ways.

They can help farmers decide what they really want, and what relative importance they place on their different goals. They might use the 'counselling' method discussed in Section 6.3 to achieve these objectives.

They also can help farmers clarify their goals and the means they are willing and able to use to reach them. So the extension agent can help the farmer on the pathways towards both knowledge and choice. It is important to note here that the extension agent might know the best information on which to act, but only the farmer will know which is the best ultimate choice.

5.5.6 Solutions

It is very important to understand the causes of a problem before making a decision on how to solve it; hence a proper diagnosis is required. As we stated earlier, the cause of the problem may stem in part from a changed situation. Research findings often give us useful insight into these circumstances. Population pressure might have forced a farmer to grow a crop more frequently in the same field. This in turn caused the eelworm population in the soil to increase, consequently reducing crop yields.

We have stressed already how extension agents can play an important role in passing on relevant research information. They may not always be prepared to wait for complete agreement among research workers about the causes of a problem. Sometimes they must make certain initial assumptions about the nature of the problem, then change their opinion later.

A farmer who makes a decision seldom knows exactly what the outcome will be. He or she might know that experiments have shown that, on average, addition of 100 kg of nitrogen per ha of rice will increase the yield by 500 kg. But whether the farmer will get the same increase this year will depend on the weather, plant diseases, soil type, etc. The extension agent has an important duty to decrease this uncertainty, by helping to make a proper diagnosis, and the farmer as an entrepreneur must decide which risks to take. It is rational for a poor farmer to be very hesitant about taking any unnecessary risk, even if it might yield him a big profit.

It is difficult for an extension agent to decide how much information to give. A farmer may be confused if given too many alternatives from which to choose, and if all the consequences of choosing these alternatives were explained. On the other hand, valuable ideas may be withheld by giving the farmer too little information. The right amount to give will depend on the farmer's capability and on the importance of the problem. It may be best to break the decision-making process for an important problem into a number of successive steps which can be solved with limited information. For example, the following steps could be taken for increasing farm size:

(1) Buy more land or intensify land use.
(2) Intensify through fishponds or through horticulture.
(3) Decide which horticultural crop to plant.

The problem here is that information given for decision 3 may make it necessary to reconsider decision 1.

5.5.7 Feedback

For good decision-making we should know whether we have achieved or improved on the desired situation. An extension agent who is helping farmers to achieve higher crop yields can improve the farmers' control over their decision-making process by teaching them how to measure crop yields. This form of control or 'feedback' should be carried out quickly. The sooner the farmers realize they cannot achieve a desired situation, the more likely they can make a satisfactory correction.

Accuracy of feedback for control also is important. Animal breeders use scales as important control mechanisms in their breeding programmes. They can decide which animals are the best converters of feed into flesh by measuring feed and water consumption and the weight gains of their animals, and thus can select for future breeding.

One of the problems we encounter with feedback information is that it is much easier to check whether some goals have been reached than others. It is easier for farmers to measure their yields than their soil fertility. This might stimulate them to use production techniques which increase their yields in the short run, but decrease them in the long run through soil erosion or depletion of soil fertility. This is no reason to avoid using the available feedback, but one should be aware of its limitations.

Table 5.1 presents a balance sheet on which the results expected from a range of possible problem solutions can be presented. An extension agent can decrease the probability that important results will be overlooked by helping a farmer to fill out such a sheet. A farmer may decide that some of these consequences shown in Table 5.1 are not important to him or her, because he

Table 5.1 The balance sheet grid

Alternative No.		
	Positive anticipations +	Negative anticipations –
(1) Tangible gains + and losses – for *self*		
(2) Tangible gains + and losses – for *others*		
(3) Self-approval + or self-disapproval –		
(4) Social approval + or disapproval –		

Source: Janis, I.L. & Mann, L. (1977) *Decision-making*, pp. 405–9 Free Press, New York.

or she does not care about social approval of others, for instance. This reduces the need to think about Point 4.

5.5.8 Collective decision-making

So far we have discussed individual decision-making, but collective decisions are increasingly important in agricultural development, for example on soil erosion control, integrated pest management, cooperatives for input supply, credit and product marketing. Much of what we have said so far is also valid for collective decisions, but in these decisions there are often conflicts of goals and interests between the different people involved. Negotiation is required to decide whose interests will be given most weight, a process in which power and negotiation skills are important. The extension agent can try to influence this process, for example, by trying to prevent development of a conflict in which each participant loses. Such conflict becomes less likely when the people involved trust each other. For this reason we pay considerable attention to these discussions in Section 6.2.3.

The whole family has a stake in many agricultural decisions, and different family members may have conflicting interests. These decisions can be made in a process of formal collective decision-making, but often are made informally when ideas are exchanged during the frequent contacts within a family. It is also possible that one member, usually the father, takes the decision on behalf of the whole family. He is not always fully aware of the consequences of this decision for his wife and his children. An extension agent can try to facilitate such a situation so that their information is fully used in the decision-making process, and their interests are taken into account. For example, he or she may try to prevent serious conflicts over division of the inheritance or division of labour and income.

5.5.9 Section summary

Different research groups approach the decision-making process in different ways. Those who follow the normative approach specify the sequence of phases you must follow before you can make a good decision. The empirical researchers argue that problems are too complicated or too variable to solve in this way. They believe that people are forced to work in a simpler fashion, by deviating only slightly from the existing situation rather than by taking all alternatives into consideration, for example. When making a choice, Bos says we must pay as much attention to our knowledge and interpretation of the facts as to our goals and means.

We have tried to integrate these different approaches to decision-making in this section by illustrating the difficulties which arise when making a good decision, and by showing ways in which an extension agent can help overcome these difficulties. The latter points are reviewed in Table 5.2.

Table 5.2 The farmer's decision-making process and how the extension agent can help

A farmer's decision-making process	Tasks of the extension agent in providing help
(1) Perception of the problem. Assessing the present situation.	Creating a timely awareness of the problem, where necessary. Increase objectivity in problem identification. Focus attention on important aspects that are amenable to change.
(2) Ascertaining the desired situation.	Consciously and systematically help the farmer to consider goals and their relative importance. Increase this insight into the manner and extent to which goals influence and possibly conflict with each other.
(3) Assessing the problem.	Objective and concrete statement of the problem. Assist with the proper diagnosis of the causes of the problem and of the barriers to achieving the desired situation. Clarify what is and what is not known of the problem. Help farmer become aware of role that emotions and socio-psychological values play in the problem.
(4) Formulating possible alternative solutions.	Expand the several alternatives farmer is considering, without giving too many. Clarify these alternatives.
(5) Evaluating alternatives.	Use scientific research results to evaluate alternatives objectively. Compare the expected results systematically with the desired situation using judgement criteria related to the farmer's goals. Overcome the difficulties of predicting the outcomes.
(6) Choosing an alternative.	Stimulate the farmer to make a choice even when uncertain, if necessary, perhaps even not to change; experiment with some alternatives. Help farmer evaluate the experiments.
(7) Implementing the choice.	Stimulate the farmer to implement the choice. Help with the decisions that are necessary for implementation.
(8) Evaluating.	Systematically and objectively check whether the desired results have been achieved. Help make results observable, for example, by measuring the harvest. Ascertain if there have been any side effects.
(9) If the desired situation has not been achieved.	Start again with a new problem. Help with decision-making in relation to this problem.

☞ Discussion questions

1 Think about how you decided to buy an expensive article, how you chose your study programme, or some similar important decision. Which functions can you identify in the decision process? In what order did they occur? Did anyone else help you? If so, in what way did they help?
2 Imagine that you are considering buying an expensive item like a wrist watch, a bicycle or a radio for the first time. What must an adviser do if he or she wants to help you make the choice as rationally as possible? Do you use one or two decision-making processes? If two, how do these relate to one another?
3 A farmer does not have enough feed for his or her animals and asks an extension agent how he or she should solve this problem. How can the extension agent help this farmer with the decision-making process?
4 The status of a farmer in a community is influenced by the number of cows owned. One farmer is considering whether to keep two cows or one genetically improved cow, because the farmer can expect the latter to give a higher income. How can an extension agent help this farmer to make a decision?

5.6 Adoption and diffusion of innovations

In Section 2.5 we discussed how extension serves as a link between scientific research and the farmer. Innovations often are developed in this research, sometimes by farmers. *An innovation is an idea, method or object which is regarded as new by an individual,* but which is not always the result of recent research. The metric system is still an innovation for some Anglo Saxon North Americans despite the fact that it was developed 200 years ago.

More than 3000 research reports have been published about the communication process involved in diffusion or spread of innovations, and about how extension clients decide whether to adopt or reject these innovations. Much of the research was carried out in US agriculture around 1950, because extension administrators were worried about delays in farmers' use of research findings. There was a boom in this research in less industrialized countries during the 1960s, because the ministries of agriculture saw the need for large numbers of farmers to use the results of scientific agriculture in order to prevent famine. People wanted to know how the adoption of relevant innovations could be accelerated.

Interviews with potential users of the innovation generally form the basis of diffusion studies. The following questions are investigated:

(1) What decision-making pathways do individuals follow when considering whether or not to adopt an innovation? Which sources of information are important?

(2) What are the differences between people who adopt innovations quickly or slowly?
(3) How do characteristics of innovations affect the rate of adoption?
(4) How do potential users communicate among themselves about these innovations? Who plays the important role of opinion leader in this communication process?
(5) How does an innovation diffuse through a society over time?

It is almost invariably assumed to be desirable for everyone in the group being studied by diffusion researchers to accept the innovation, a reasonable assumption for many innovations, but certainly not for all. Mass vaccination of cattle against foot-and-mouth disease is a good example. Few farmers would object on principle to a campaign which protected their cattle from this disease, particularly if veterinarians vaccinated their animals without charge. Extension agents often will be more effective in diffusing these innovations when they base their strategies and methods on adoption and diffusion of research findings. However, this research does not provide much help to the extension agent who wishes to improve farmers' management decision-making skills for achieving their goals in the specific situation of their enterprises. There are many situations in which all farmers cannot be recommended to adopt an innovation because this decision should depend on their resources and personal values, for example, a change from subsistence farming to commercial vegetable production.

5.6.1 Adoption process

Research studies have demonstrated clearly the extensive delays which often occur between the time farmers first hear about favourable innovations and the time they adopt them. It took four years on average for the majority of mid-Western US farmers to adopt recommended practices. Research workers naturally have been keen to find out what happens during this time. The following stages, which resemble the normative decision-making model (Section 5.5), often are used to analyse the adoption process:

(1) Awareness: first hear about the innovation;
(2) Interest: seek further information about it;
(3) Evaluation: weigh up the advantages and disadvantages of using it;
(4) Trial: test the innovation on a small scale for yourself;
(5) Adoption: apply the innovation on a large scale in preference to old methods.

The adoption process does not always follow this sequence in practice. For example, it is not feasible to test a new farm building on a small scale. Interest may precede awareness when farmers are looking for a method to control what for them is a new and unknown crop disease. Some authors place another awareness stage before stage 1, a general openness for new information such as subscribing to a farm paper, which makes farmers aware of innovations.

There is insufficient evidence to prove that these stages exist. Decisions in practice often may be made in a much less rational and systematic manner than we have outlined here. In the latest edition of his book *Diffusion of Innovations*, Rogers (13) proposes different stages in the process:

(1) knowledge;
(2) persuasion (forming and changing attitudes);
(3) decision (adoption or rejection);
(4) implementation;
(5) confirmation.

He indicates there is clear evidence for the 'knowledge' and 'decision' stages, but evidence for the other stages is much less certain. Perhaps persuasion and implementation can happen at different moments in the adoption process. Persuasion can occur after the decision to adopt, which sometimes is taken without careful consideration of the possible consequences. Implementation, which is a serious consideration of how the farmers will change their farm management by adopting this innovation, can take place partly before the decision is taken. Implementation often implies that the innovation is modified to suit more closely the needs of the farmer who adopts it. People often gather additional information after they have adopted an innovation to confirm they have made the right decision.

Another important difference between the old and new set of stages is that explicit attention now is given to the possibility of the innovation being rejected.

This change in the adoption model is related to our previous discussion of decision-making. The normative decision-making model no longer is considered to provide an adequate explanation of the way people make decisions. This also made it necessary to abandon the idea of stages in the adoption process.

The implementation of innovations has received more attention in recent years. After a farmer has decided to adopt vegetable growing, for example, the implementation of this decision requires considerable additional learning and decision-making in how to grow these vegetables most effectively. In this, and in many other cases, we are not dealing with the adoption of one innovation, but with a package of innovations. Innovations often must be adapted to the specific situation in which they will be used.

Extension research has shown that different sources of information are important for first hearing about an innovation and for making the final decision to adopt or reject it. In countries with a well developed mass media system farmers usually get the first information on innovations through these media. But they first like to discuss it with somebody in whose competence and motivation they have confidence before they decide to adopt this innovation. This person may be an extension agent, but for most farmers a discussion with one of their colleagues and/or an observation of the colleague's experience is more important. In countries whose mass media system is not yet well developed, demonstrations often play an important role. Dasgupta's overview of 300 studies in India (14) shows that extension agents there are mainly influential during the early stages of the adoption process.

5.6.2 Adopter categories

It is understandable that not everyone adopts innovations at the same rate. Some people accept new ideas years before others. Differences between people who readily adopt innovations and those who play a waiting game are an interesting topic for investigation. Many of these investigations combine a sample of several innovations into an 'adoption index', rather than studying a single innovation. Innovations commonly studied often relate to economic growth and are based on scientific research, for example, agricultural methods which increase yield per hectare or per person.

The adoption index usually is calculated by asking people if they have adopted any of 10 to 15 innovations recommended by the local extension service. They receive a point for each one adopted. A difficulty is that there are often very good reasons why a person cannot adopt a certain innovation. Use of a particular machine on large farms may indicate how modern a farmer is. However, use of this machine by small farmers may be an indication that they have not worked out very carefully how profitable the machine is for them. Hence, if an adoption index is used, it should be based on the percentage of innovations adopted which are applicable to a given situation.

People often are divided into five categories according to their scores on an adoption index. These are:

(1)	Innovators	2.5%
(2)	Early adopters	13.5%
(3)	Early majority	34.0%
(4)	Late majority	34.0%
(5)	Laggards	16.0%

These percentages serve mainly to make findings from different studies comparable by using a uniform classification.

Classification of people in these different adopter categories by definition depends on the degree to which the whole group has adopted the innovations, and on the assumption that distribution of adoption over time is normal. Borderlines between categories for this classification are drawn at one or two standard deviations (sigma) from the mean, as is shown in Figure 5.10.

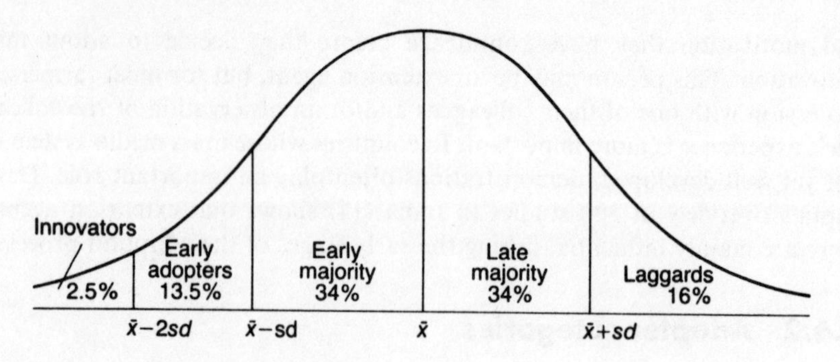

Figure 5.10 Adopter categories
From: Rogers, E.M. (1983) *Diffusion of Innovations*, 3rd edn., p. 247. Free Press, New York.

Many research workers have investigated the relationship between an individual's adoption index and a variety of his or her social characteristics. Variables which are related to the adoption index have been investigated in such diverse areas as agriculture in industrialized and less industrialized countries, education, health services, and consumer behaviour. Remarkably similar results have been found in all of these fields. It appears to make little difference whether you are an American doctor or an Indian farmer. Some of the results are summarized in Table 5.3.

Table 5.3 Percentage of studies showing positive relationship between adoption index and other variables

Variable	% of studies	No. of studies
Education	74	275
Literacy	63	38
Higher social status	68	402
Larger size units	67	227
Commercial economic orientation	71	28
More favourable attitude to credit	76	25
More favourable attitude to change	75	57
More favourable attitude to education	81	31
Intelligence	100	5
Social participation	73	149
Cosmopolitanism (urban contacts)	76	174
Change agent contact	87	156
Mass media exposure	69	116
Exposure to interpersonal channels	77	60
More active information seeking	86	14
Knowledge of innovations	76	55
Opinion leadership	76	55

Based on: Rogers, E.M. (1983) *Diffusion of Innovations*, 3rd edn., pp. 260–261. Free Press, New York.

Care should be taken in interpreting this table, because it does not distinguish between cause and effect. Many studies show clearly that people who have adopted many innovations have frequent contact with change agents. Is that because this contact results in the adoption of innovations, because people interested in innovation seek contact with change agents or because change agents seek contact with these people? Probably all three factors play some role. Also, some of the relationships, while being positive are very low.

Age is not included in this study, because, contrary to popular opinion, no relationship was found in about half of the studies, and only one third of the other studies showed that younger people are more innovative than older people.

These investigations allow us to predict which type of people will apply extension advice quickly in practice. We can conclude that often, but not always, the same relationship was found in these studies. New studies, such as those investigating the relationship between adoption and education level for example, are not so necessary now. It would be more useful to know why we found these relationships.

Most of the studies summarized in Table 5.3 are based on innovations developed by research. Innovations not based on research may be adopted by different kinds of people. For example, some US farmers are interested in 'organic farming' based on low use of chemical inputs. Innovators for this type of farming often are farmers with small farm size and low levels of education (15).

For a long time we thought that people adopted innovations only very slowly because of their traditional or conservative attitude towards life. We call this the 'individual-blame' hypothesis. Research among Latin American farmers has focused attention on the 'system-blame' hypothesis which states that it is not sensible for farmers to adopt ideas in their present situation. Either they do not have sufficient resources, or the power relationships in the society are such that estate owners, money lenders, traders and others profit from these innovations rather than the farmers themselves. It also is possible that innovations are not available in remote villages or inputs are sold only in much larger quantities than a small farmer can use and afford. The reason that small farmers do not use fertilizers might not be their traditional attitude towards this innovation (individual-blame), but the fact that they have to borrow money for it at a high interest rate, that they get only a small return when they sell the additional crop because they are in debt to the merchant, that fertilizers are not sold in their village and/or that they cannot afford the risk that they might not get any yield increase by applying fertilizer in a dry year (system-blame). It does not seem sensible to us to state that either the individual-blame or system-blame hypothesis is correct. Each concrete case should be tested with both hypotheses. Both hypotheses can contribute to clarifying reality, despite the fact that the accent will lie more with one than the other (16).

We must realize that the correlation between adoption of innovations and income or capital is much lower than one. There are some poor farmers who adopt innovations rapidly. We often encounter this situation when surveying farmers in one community at one point in time, and even more commonly when studying farmers during a generation. Some farm families who used to be poor are now relatively wealthy. The number to have become wealthy depends on the rigidity of the social structure. There will be many more in a rapidly changing industrial society where many new opportunities open up than in a traditional part of India with its caste structure. In other countries there might be other new opportunities, such as fish ponds. It is also possible that the new opportunities require substantial capital for the modern technology developed in industrialized countries, as with poultry, for example. Therefore, only wealthy enterprises might be able to use them.

5.6.3 Innovations

Adoption and diffusion research has usually studied innovations in production technologies, but there are also very different kinds of innovations to which extension agents should give attention, such as:

(1) New methods to assist management decisions, such as soil testing, linear programming or computerized expert systems (see Section 6.6). Teaching farmers to use these methods increases their skills for making decisions about the specific situation of their farm, and can make them less dependent on extension agents.
(2) New farming systems, such as a change from crop production to commercial horticultural or animal production.
(3) New social organizations such as farmers' unions and cooperatives.

A new technological innovation always has two components – the hardware and the software. The situation is clear in the case of the computer where the machine (hardware) is useless without the programs which instruct it what to do (software). It is also true for a plant variety where we have the plants (equivalent to hardware) and the techniques for growing them (equivalent to software). It is possible that the same software used for an old variety can be used for a new variety, but we often need new techniques for cultivation, fertilizer application, etc. to ensure optimal production by the new variety. Farmers often play an important role in developing the right kind of software. Scientists who develop new hardware in their research station should take into account the access farmers have to resources used in developing and using their software. However, scientists often fail to do this. For example, they have developed many more techniques for irrigated agriculture (which could include software) than for rain-fed agriculture, although the vast majority of farmers do not have irrigation.

Farmers also play a major role in the development of new farming systems. Elements of a system can be studied at a research station, such as testing the effect of different feeds on crossbred animals, for example. However, changing from local low producing cattle to crossbred cows for commercial milk production also requires changes in the production of animal feed, the integration of crop and animal production, labour requirements at different times of the year and product marketing. These can only be studied in the actual farm situation. Hence the farmers who are innovators in introducing such a new farming system are an important source of information for their colleagues. Stimulation of communication between the innovators and other farmers then becomes a major role of extension agents. They may do this by organizing farm visits and meetings where these farmers tell about their experiences, or perhaps by passing on what they have learned from the innovators. It requires quite a different extension approach from the traditional transfer of technologies because changes in a farming system involve a lot of risk which only the farm family can decide whether they are willing to accept.

The first farmers who adopt or develop a new farming system are often quite different from those who are innovators in production technologies. In the Netherlands, for example, it was mainly the small and poor farmers who started horticultural, poultry and pig production farming systems which require a lot of labour. Many of their grandchildren now have much larger farms than the grandchildren of the arable farmers who had the large and high status farms a century ago. Market demand for the horticultural, poultry and pig products has increased much more than demand for arable crops. The choice of the optimal farming system often has more impact on farm income than the choice of the optimal production technology. The extension service should play an important role by helping farmers make this choice rather than by telling them what they should do. In this way extension agents help farmers to discover and analyse how new opportunities can be explored, and to predict the consequences of their adoption. The farmer is responsible for making the choice whether or not to take the risk. In our view it is quite acceptable to first give help to farmers who are relatively well-to-do, because they manage their farms very well. In this way the extension agent learns perhaps more than the farmer and can use this new knowledge to help other farmers later.

New social organizations can only be developed by the farmers themselves. Attempts by governments to tell farmers how they should organize themselves have usually failed. Farmers know much better than government officers which kind of organization can work in their culture and their social structure, and how such an organization can be controlled by its members to prevent misuse of funds. Extension agents can stimulate farmers to think about the best way to organize themselves by informing them of the experiences, successes and failures with farmers' organizations elsewhere.

A number of studies have analysed the relationship between characteristics of an innovation in production technology and its rate of adoption. Most have used more or less objective judges, or have assumed that all farmers perceive these characteristics in the same way. This may be why all studies do not come to the same conclusion. However, they do indicate that the following characteristics are important:

Relative advantage

Does the innovation enable the farmer to achieve goals better or at lower cost than he or she could previously? This advantage can be influenced by giving incentives to the farmer, such as by providing seeds at subsidized rates. Such incentives may motivate farmers to try the innovation. However, it is often difficult for a farmer to perceive advantages caused by a change in probabilities. For example, an inoculation may reduce the probability that his or her animals will contract a certain disease. Researchers often study the yield per ha in order to determine this advantage although yield per man-day in a busy period is more important in some farming systems.

Compatibility

Compatibility with socio-cultural values and beliefs, with previously introduced ideas or with farmers' felt needs is important. Clearly, it is very difficult to introduce pig husbandry among Moslems even if it is a very profitable enterprise. On the other hand, farmers who have received large yield increases by growing improved wheat varieties are likely to be very happy and to accept improved rice varieties. However, if an innovation has failed after introduction, it will be very difficult to get similar innovations adopted.

Changes also can be complicated, because the farmers' environment, as well as the farmers themselves, often have to change, for example to provide them with necessary inputs and markets.

Complexity

Innovations often fail because they are not implemented correctly. Some require complex knowledge or skills. For example, it may be necessary to introduce a package of several relatively simple but related innovations. Each on its own is easy, but the relationships between them may be difficult to understand. Dairy cows with higher genetic potential will produce more milk only if they have feed which is higher in protein and energy content. This in turn will require different crop husbandry practices. Otherwise the 'improved' cattle will produce less than indigenous cattle, as has often been the case with cattle imported to less industrialized countries through aid projects.

Trialability

Farmers will be more inclined to adopt an innovation which they have tried first on a small scale on their own farm, and which performed better than an innovation they had to adopt immediately on a large scale. The latter involves too much risk. Trialability may be related to *divisibility*, as with fertilizers, for example. Although large machines cannot be 'divided', sometimes they can be rented before they are purchased.

Agricultural development programmes also may increase the trialability of innovations. For example, before constructing a large dam for irrigation it might be useful first to construct a number of small tanks which can be used to help farmers learn how to change from rain-fed to irrigated agriculture. This would involve extra cost, but that might be saved by more rapid utilization of the full potential of irrigation water after the dam is constructed.

Observability

A farmer can see a mile away that a colleague has changed from fodder beets to maize as cattle feed, but the farmer might not know the bookkeeping system used by a nextdoor neighbour. A farmer might not show his or her improved cattle to neighbours out of fear of the 'evil eye'. Farmers learn much from observing and discussing their colleagues' experiences, their observations often being a reason to start discussions.

Extension agents who still have to gain the confidence of their farmers should start by promoting successful innovations. Hence they should look for those which diffuse rapidly. However, after a while they might spend most of their time on those innovations which have a considerable impact on farmers' incomes or goal achievement, even if they would not diffuse without their efforts. There are also innovations which diffuse rapidly without much attention from extension organizations.

Market research in business pays considerable attention to consumers' needs for and perceptions of new products (17). This information is used to develop products which are likely to be a commercial success, and to develop effective communication strategies for marketing the products. Marketing managers would not know which messages to stress in advertising and other sales promotion activities without such research. Choice of the right name for a product also requires substantial testing, because the name can influence perceptions of the product.

This type of research is less common in agriculture and extension than in business, partly because government officers will not lose their jobs if they do not 'market' their ideas effectively. Furthermore, there are seldom adequate resources for good market research. However, an effective flow of information from field extension agents to extension managers and research workers can make up for much of this deficiency.

It is difficult to know when to release an innovation for farmers to use. It may fail if released too soon. What might the consequences of failure be? Will it mean just a small decrease in yield or that some members of the family starve? If we release it too late we miss some of its advantages and farmers may be unable to compete with those in another country who adopted the innovation earlier.

5.6.4 Diffusion processes

We said in Section 5.6.1 that adoption of innovations is influenced strongly by members of social groups. This means that when some members of a group have adopted an innovation others often will follow. We then must ask how extension agents can use and influence this diffusion process. Hence, we have to know who exercises influence in these social groups.

Farmers are keen observers of other farmers' activities, and in many countries, but not in all, they spend much time discussing their farm experiences with friends and neighbours. They learn much in this way, but they learn more from some colleagues than from others. They know who gets good yields or good results in their village, and who experiments with new methods. Some of these successful or progressive farmers are willing to share their experiences with other farmers. In this way they become opinion leaders in the village because they help other farmers solve problems considered to be important. Thus, opinion leaders have considerable influence on the way in which people in their village think and farm. As extension agents are not able to work closely with all farmers they should concentrate their attention on these opinion leaders, for example, by choosing them as contact farmers in the Training and Visit System (see Section 10.9).

The problems considered to be important depend on the norms of the group. Those farmers who are able and willing to help their colleagues with these problems usually will become opinion leaders in the group.

An opinion leader will fulfil several of the following functions in his or her group with regard to innovations:

- passes on information from outside the group;
- interprets outside information on the basis of his or her own opinions and experience;
- sets an example for others to follow;
- 'legitimizes' or rejects changes that others want to carry out. That is to say, the opinion leader gives his or her approval or disapproval for these changes; and
- is influential in changing group norms. Not all opinion leaders will do all of these things. There may be some opinion leaders who provide information early in the adoption process, and others who legitimize the decision to adopt or reject an innovation.

Studies generally found that opinion leaders:

- adopt many innovations, but usually are not the first to adopt them;
- are well educated and enjoy sound financial positions in their communities;
- lead an active social life and have many contacts outside their immediate surroundings; and
- have a special interest in their subject.

Interaction between members of the different social strata can be rather limited in communities with rigid social stratification. Therefore, each of these strata can have its own opinion leaders who have limited influence among other strata, as is the situation among large landowners and peasants in Latin America. The extension agent may have to look for opinion leaders in each of the social strata.

People who are opinion leaders about specific issues also may be opinion leaders in a wider context, especially in traditional societies. However, opinion leaders tend to be more specialized in modern industrial societies; someone who leads political opinion may not be so respected for his or her opinions about plant protection. Many innovations can only diffuse effectively among farmers who are more or less homogenous in resources and social status. Some innovations which are suitable for one group are unsuitable for another group, especially where there are large differences in farm size, or some farmers have irrigated land and others have to rely on rain-fed agriculture. Furthermore, social interaction between these groups often is limited. Extension agents in these situations should try to establish contacts with opinion leaders from each of the groups. In many countries there are also separate communication networks for men and women. Information given by extension agents to male opinion leaders often will reach farm women very slowly and be rather distorted, if it reaches them at all.

Opinion leaders are members of small social groups who influence other members of their group. Formal leaders, such as a village headman or tribal chief, and national idols, such as film stars, pop musicians, and some religious leaders and politicians, also can have considerable influence.

Formal leaders have power by virtue of their position, but also clearly have been able to solve many problems faced by their people. Hence, other villagers may follow their advice or example in the belief they will be successful also. These leaders may exert influence on a wide range of topics in traditional societies. They also may have considerable power over collective decisions such as irrigation or grazing rights.

People tend to accept new ideas most easily from others who are similar in many respects, like a brother, for example. Outsiders are not trusted, especially by traditional villagers. The problem is that close friends and relatives are likely to have fewer new and valuable ideas than an outsider. Hence, farmers from other districts can be very helpful in extension work because local farmers are more likely to accept their ideas than those of extension agents or other government employees. On excursions to these districts it is important for farmers to be given the opportunity to have adequate informal discussions with their hosts after they have demonstrated innovations adopted successfully.

Opinion leaders

Extension agents will not be able to work closely with all farmers in their district, so they can increase their impact by cooperating with the opinion leaders. They can influence a large proportion of the farmers by choosing opinion leaders from among those they work most closely with. This is relatively easy in a community where group norms favour the adoption of innovations, as we can expect the opinion leaders also to favour cooperation with the extension agent. There will be some innovators and early adopters in these villages who are not opinion leaders. Extension agents should avoid spending too much of their time working with these farmers if they wish to influence the whole community.

It is more difficult working in communities where most farmers are not yet interested in adopting innovations. Farmers in these areas who show interest in cooperating with the extension agent usually are not the opinion leaders. In this situation the extension agent should try to gain the opinion leaders' trust and develop their interest in modernizing agriculture. Then in turn he or she can influence other farmers. This must be done carefully because the opinion leaders will lose their leadership position if they change too quickly. It is also somewhat risky betting on which young farmers will be leaders ten years from now.

The extension agent should be aware that there may be conflicts between different religious, tribal or status groups in the community. For example, small and poor farmers may be convinced that the large and high status farmers are trying to exploit them. Thus the extension agent may lose the trust of one faction by cooperating with another. He or she should try to cooperate with all factions, although this is much easier said than done where there are serious conflicts.

How can extension agents identify opinion leaders with whom they could cooperate? In the first instance they could observe carefully the social processes in the community. People will react in a different way according to who is speaking, say, at a meeting or in informal discussion. Farmers will refer to what they have seen on other farms or have heard from other farmers, or in a discussion with the extension agent. Agents will find it useful to keep notes of these pointers to leadership structure in the community.

Extension agents also could seek assistance from some villagers or outsiders who know the community well. For example, they can point out in the village council that a demonstration will be more effective if it is held on an opinion leader's farm, then seek suggestions about where it should be held. The danger with this approach is that their advisers might mention only people from their faction if there are conflicts in the area. It is also feasible to make a list of all local farmers and ask the advisers to judge each farmer according to the influence he or she has in discussions about farming. However, if the opinion leaders change rapidly through these contacts they might lose their leadership position because they deviate too much from group norms.

The most reliable method, but also the most time consuming, involves sociometric analysis. Each farmer is asked questions such as 'Could you give the names of two farmers with whom you talk most frequently?' or 'From which farmers would you seek advice about whether or not a new method will work well on your farm?'. Responses to such questions make it possible to count how often each farmer in a village is chosen. Those chosen most frequently are considered to be opinion leaders. The line dividing leaders from followers is drawn arbitrarily. There are important differences between cultural groups in their willingness to share information and experiences and to learn from each other. Agricultural development usually is fastest among those groups where this willingness is high, because the intelligence of all farmers in this situation contributes to development. The extension agent plays an important role in stimulating interaction about innovations among farmers. The Farmer Field Schools we discussed in Section 5.3 and study clubs in which farmers discuss and analyse their experiences can play an important part in this process.

5.6.5 Some implications of adoption research for extension (18)

Researchers who study the adoption and diffusion of innovations analyse how adoption processes happen at that time. They do not describe how things happen in every community, but how things might happen. They can help draw an extension agent's attention to some processes and variables which probably are important among his farmers. We may use such research findings either to accelerate or to change adoption processes. Attention has been given to this latter point only recently in relation to interest in the system-blame hypothesis. This leads people to wonder whether it is possible to change the system.

We will discuss first some possibilities for accelerating adoption. Research has shown that different information sources are used at the beginning and the end of the rather lengthy process of adopting an innovation. Hence we should combine different media so they reinforce each other in an extension programme. The mass media can be very useful in creating interest in interpersonal communication with other actors in the AKIS about innovations. However, they are relatively ineffective in changing behaviour unless there is extensive personal follow-up by extension agents in the field.

Choice of farmers who participate in group meetings and who are visited by extension agents also is very important. If the farmers choose these people themselves, most of the contacts are likely to be with the innovators and the early adopters. The extension agent can try to establish contacts with the opinion leaders in order to increase his or her impact on a wider group of farmers.

Active promotion of an innovation may be taken over from extension agents by farmers who have adopted it already. Such farmers are not always well suited to this task if the innovation is difficult to implement, for example, keeping and analysing accounts. Farmers in the Indonesian Integrated Pest Management programmes are used successfully as facilitators in the Farm Field Schools. The innovation also may be suited only to a certain category of farmer, such as those with large farms. Income differences between farmers may increase if there is no further promotion of the innovation after the early adopters have accepted it, which may not be a desirable outcome of the extension policy.

Considerable attention has been paid to accelerating adoption among those farmers who are slow to adopt. This is because these farmers tend to be poor, and many people believe it is desirable to decrease income differences among farmers, or in other words to increase equity.

Research findings about the characteristics of different adopter categories provide extension agents with important information for choosing their target groups. It may be necessary to develop different messages to meet the needs and situation of each category of adopters. This approach frequently is used in marketing where it is called *market segmentation* (compare points raised in Section 7.3.2). Different extension methods may be required for different categories, for example because women do not attend meetings or because only large farmers have a television set.

Farmers who regularly seek advice from extension agents appear to be relatively well informed through many other communication channels. They read farm magazines and trade journals, attend field days, demonstrations and meetings, discuss innovations with people who have experience outside their immediate district, and so on. If extension agents try to answer their requests for advice, this has the following advantages:

- these farmers are interested in extension advice, so there is a relatively high probability they will adopt innovations recommended by extension agents;
- the extension agents gain experience with the innovations in their own environment. Innovators play an important role in testing and adapting these innovations at the local level;
- relatively rapid expansion of production is achieved because the innovations tend to be adopted on large farms;
- solutions to the problems of these large farmers often have been provided

by agricultural research, but such research often has little to offer to small farmers who seldom ask for advice; and

- it is difficult for extension agents to refuse large farmers' requests for help. They often have extensive influence with other farmers, and sometimes with more senior staff of the extension service.

Extension agents who answer all requests for advice and information will have little time to assist farmers who do not ask for help. These farmers often have serious problems, but because they are only partly aware of them, they may not be able to ask accurate and well thought out questions. Leaving the choice of a target group to those who ask questions can lead to even greater differences in income among farmers. Those who adopt good innovations first generally profit most from them. They also run the risk that new innovations will not prove to be as good as they were first thought to be.

More extension agents now believe it is essential to contact farmers who seldom ask for help. Despite his best intentions, an agent may not be able to help poor farmers. Research may have found some profitable solutions for large, well-capitalized and irrigated farms, but none for poor farmers working under rain-fed conditions. Extension agents and policy makers may have to redirect agricultural research towards solving the problems of these small farmers. In the meantime, extension agents can offer limited help by noting why some farmers get higher yields than others, then developing messages for the less successful on the basis of what they have learned from the more successful.

Extension agents also may be able to help poor farmers to organize themselves into groups which help each other, and which have more power in dealings with salesmen, moneylenders and/or landowners. Such a move can be an important way of changing the system which currently makes it unprofitable for poor farmers to adopt innovations, although it may also arouse resistance from those who fear they may lose some of their power. It is sometimes possible to convince them that everyone will be better off with increased productivity.

5.6.6 Limitations of adoption and diffusion research

Diffusion of innovations is the area of extension education which has received the most support from empirical research. Unfortunately, many diffusion research workers want to keep digging deeper into questions asked by their predecessors. We should be asking new questions and studying new problems if we wish to advance science and solve practical extension problems. We illustrate these points with the following examples:

Scientific knowledge

Adoption and diffusion research studied certain aspects of the utilization of scientific knowledge, but ignored the way this knowledge is produced and how it is integrated into ways in which the farmer sees his own problems. It might be useful also to study who decides why certain innovations are developed, and which of these are diffused to a certain group of farmers. For example, it can have significant influence on the process of agricultural development whether the adoption of labour saving or of yield increasing innovations are stimulated. The power structure in the society may have much influence on these decisions.

Diffusion research usually assumes that all innovations originate at research institutes, whereas in fact many originate either from farmers or are modified by them to adapt the innovations to their situation. This research also assumes there is a centralized agency which tries to diffuse innovations, whereas some innovations are diffused mainly by a decentralized system of active farmers, NGOs and traders.

The desirableness of innovations

Much diffusion research assumes implicitly, and we believe incorrectly, that the adoption of innovations always is desirable. The nature of the innovation and the circumstances and goals of the farmer must be given strong consideration first. Undoubtedly mechanization and large scale farm methods would increase labour productivity in many developing countries. But the side effects would be little short of disastrous in many ways. Already critical levels of unemployment would increase because less farm labour would be required, important reserves of foreign currency would be used to buy the machinery, and so on. Fortunately, most rural development workers now recognize agricultural production as part of a wider system in which changes in one component of the system may have unexpected consequences for other components. There is now less uncritical acceptance of innovations, and more attention is paid to the development and use of *appropriate technology*, or ideas and methods suited to the needs and conditions of the majority. Most diffusion studies have assumed implicitly that a farmer's major goal is to increase income. However, risk aversion can be a more important goal among poor farmers, especially in less industrialized countries. We might reach different conclusions if we studied innovations which help farmers reach this latter goal. For example, it may be that the categories of farmers who are the first to decide to grow different varieties of bananas to decrease their risks differ from the farmers described in Section 5.6.2.

Advantages and disadvantages to the user

Few diffusion research projects have looked at the advantages and disadvantages of innovations from a potential user's viewpoint, whereas this is important information for developing innovations which solve some of the farmers' problems. Groups of unrelated innovations, rather than single ideas or methods or a package of related innovations, have been studied by compiling adoption indices. These make it impossible to find out whether farmers rejected a fertilizer because they did not think it would increase yields, because they cost too much, or because they thought it would lower soil fertility. Clever advertisers pay much attention to these questions when they formulate an effective advertisement (17). A large Australian fertilizer company had been marketing its products for years on the basis of the chemical contents, with names indicating the ratio of nitrogen, phosphorus and potash. When sales of their nitrogen fertilizers failed to respond to increased advertising, the company enlisted the help of a rural communication specialist who interviewed farmers. The marketing strategy was changed to stress the end use of the product rather than its composition. Each product was advertised on the basis of how, when and where it should be used and what results could be expected. In other words, they changed from a technical source orientation to a receiver-user orientation with their messages.

Content and interpretation of the message

The content and interpretation of messages given to farmers also require more research. The role of different information sources has been studied extensively, but little attention has been paid to how the message content is selected and treated by the source, decoded by the receivers and incorporated in their knowledge, attitudes and behaviour. The Australian fertilizer company mentioned above knew much about the radio and television stations, newspapers and pamphlets that were channels for its advertising messages, but had not looked critically at the content of these messages from their readers' viewpoint. It soon was found that farmers could not understand many of the technical points drawn from research reports, selected by technically-minded company men, and written in semi-technical language.

Application of innovations

Adoption studies generally assume sufficient research information is available for extension agents to give good advice on how to apply innovations. This may not always be the case. Whereas diffusion research has shed little light to date on how to combine research findings with the farmers' experiences and their ideas about their situation and their personal goals, Farming Systems Research focuses on these problems.

Feedback

Few systematic checks have been made of farmers' reactions to extension agents' efforts to promote diffusion and adoption. Farmers' answers on questionnaires give us relatively little information about this problem. Some attempts have been made to test different combinations of extension methods experimentally. However, reports of the results of these experiments do not discuss in detail how the methods were used.

The social effects of innovations

Most diffusion studies have concentrated on relatively small and discrete technical changes, such as replacing hand hoeing with chemical weed control. Little attention has been given to major changes in social structure or an individual's way of life, for example, when changing occupation or changing from subsistence to commercial agriculture. We have been more concerned with 'peripheral' innovations than with those which are 'central' to a social system. Social research findings offer little guidance to extension agents who have to advise farmers about such changes. This is partly because we have assumed that adoption of innovations is desirable for everyone. We will discuss this problem further in Section 6.3.

The effects of social systems

Changes by individuals and groups have been the focus of much diffusion research. Changes in institutions and societies seldom have been investigated, despite the obvious importance of social changes taking place in rural societies everywhere. The structure of the society can have much influence on the opportunity and desirability for different groups of farmers to adopt certain innovations. These system effects have not received much attention in diffusion research to date.

Most adoption research is based on individual decision-making, but collective decisions are of increasing importance for agricultural development. We often have to reconcile conflicts of interests between different actors in these collective decisions, but adoption and diffusion research offers little help with conflict resolution.

5.6.7 Section summary

The process by which many types of innovations diffuse and are adopted has been studied extensively. Many years may pass between the time people first hear about some innovations and the time they adopt them. Mass media play an important part in the early stages of the process by making people aware of the innovations. Personal contact with people who are known and trusted, such as extension agents and opinion leaders, is more important as the process progresses.

People who are quick to adopt innovations may be characterized by:

- having many contacts with extension agents and other people outside their own social group;
- active participation in many organizations;
- making intensive use of messages from the mass media, especially those which carry expert information;
- being well educated;

- having a relatively high income and standard of living;
- having a positive general attitude to change; and
- having high aspirations for themselves and for their children.

Opinion leaders play an important role in diffusing innovations. They tend to be people who are capable, willing and in a position to help others solve important problems. Who becomes an opinion leader in a group depends on group norms and current problems facing the group. Messages about innovations will be most successful when the receiver trusts the source and shares similar attitudes towards the innovation. Future research will have to pay more attention to differences in the adoption of new technologies, new methods of making management decisions, new farming systems and new social organizations.

The rate of adoption is influenced by the farmers' perception of the characteristics of the innovation, the changes this innovation requires in farm management and the roles of the farm family. Innovations usually are adopted most rapidly which:

- have a high relative advantage for the farmer;
- are compatible with the farmer's values, experiences and needs;
- are not complex;
- can be tried first on a small scale; and
- are easy to observe.

Results of adoption research can be used by extension organizations to accelerate the rate of adoption of innovations or to change adoption processes in such a way that certain categories of farmers adopt innovations more rapidly. Hence it may be necessary to develop different extension messages for different categories of farmers.

☞ Discussion questions

1 An extension agent wishes to promote the adoption of an innovation he believes will improve living conditions for all of his farmers. He decides to concentrate his attention on successful farmers who have regular contact with him. The extension agent expects his farmers to use their influence as opinion leaders to persuade others to adopt the innovation relatively quickly. What do you think about this strategy? Why?

2 How do the decision-making process and the adoption process agree with and differ from each other?

3 Many research results take many years to be applied generally in practice. What could be the reasons for this delay? How can you speed up the process?

4 The engineering section of an experiment station in a less industrialized country has developed a simple irrigation hand pump which can be made from pieces of metal and rubber available in most villages. Research workers have calculated that use of the pump, which costs the equivalent of two months' income for a small farmer, will save farmers time and effort in irrigation, which they could put to better use in growing extra food. How can an extension agent promote the adoption of this innovation in rural villages?

5.7 Chapter summary

Perception, communication, learning and decision-making are discussed in this chapter as distinct processes, yet they are clearly related. They are in fact different ways of looking at the same process. Evaluation of decision-making processes, feedback of communication processes and rewards and punishment of learning processes are basically the same processes, each of which is based on perception of the environment.

We have a cognitive map in our mind of the physical and social environment in which we live. We make our decisions on the basis of this map. On the one hand our map is based on our perception of the environment, while on the other hand it guides our perceptions. We are inclined to explain our perceptions in a way which confirms the map, unless it becomes clear that we have to revise it. Our decision will give us the results we expected if our map agrees with the real environment. Often we experience this through the feedback we get in our communication with other people. We receive a reward in this way. However, if our cognitive map deviates from our environment on important points because we perceived this environment incorrectly, we will receive negative feedback which is a form of punishment.

We learn from our own experiences in this way, but we learn also from communicating with others about their experiences. This is why other farmers are very influential as information sources in the diffusion process. Whether or not we perceive our environment and the messages we receive from others correctly depends on the one hand how clear this information is. For example, it is easier to perceive the effects of nitrogen on crop growth correctly than it is to perceive the effects of crop rotations on soil fertility. On the other hand, it depends on what the consequences for our acceptance of the message are. It is less of a blow to farmers' self-esteem to accept that they were not well informed about the latest plant protection research findings than to accept that their basic management decisions about which crops to grow were wrong.

We can accept ideas as we receive them in this communication process, but these ideas also can stimulate us to think, thus changing our cognitive map of reality. The fact that we have been able to discover new relationships between facts and phenomena will serve as a reward. We will try to test whether this new cognitive map is indeed better than the old one in our actions and in our communication with other people.

The major task of an extension agent is to help farmers to develop as accurate a cognitive map of their environment as possible. This enables them to make decisions, that is, to behave in a certain way, so that they achieve their goals as closely as possible. Also, this cognitive map is influenced by the goals themselves, especially with respect to the probabilities and the costs of reaching them.

Adoption processes are special kinds of decision-making processes regarding the adoption or rejection of an innovation. They are highly influenced by the communication we need for forming a cognitive map of the consequences we can expect from this innovation. Whenever possible, the farmers will test the innovation first on a small scale. They will have more confidence in feedback on their own decision-making than in what they learn about the feedback others have received when they tried this innovation. The other farmer or the research station which tried it might not have analysed their feedback correctly, or it might not be communicated correctly, or the situation in which the innovation has been tried might differ from that of the farmers. The latter point usually is true on a research station. Therefore, many farmers have learned to distrust recommendations based on research findings which have not been tested in their own environment. Extension agents have an important role in helping farmers correctly interpret feedback they receive from their experiences.

Humans are both rational and emotional beings. Our emotions influence our thinking, and our thinking influences our emotions and feelings. Our perceptions are influenced by our feelings towards ourselves as well as towards the objects and messages we receive. Communication processes are influenced by the sender's and receiver's feelings and attitudes towards each other, towards the message, the channel and themselves. These also influence learning processes. The motivation to learn is most favourable if the 'student' has a real interest in the subject, but is not too deeply interested. Feelings influence decision-making processes, especially those regarding problem identification and choice of goals. Feelings towards the innovation play a major role in learning from adoption processes.

Extension agents deal with farmers as whole people, with their cognitions or thoughts and with their feelings, whether the agents like it or not. Their own feelings and the way they handle them influence the farmers' reactions towards the agent and towards their message. Farmers often will be able to form sounder opinions and make better decisions if the extension agent can help them to become aware of how their feelings influence their opinion formation and decision-making.

Guide to further reading

The application to marketing of theories summarized in this chapter is discussed in:

Engel, J.F., Blackwell, R.D. and Miniard, P.W. (1993) *Consumer Behavior*, 9th edn. Dryden Press, Chicago.

Perception

For a more detailed discussion of psychological aspects of perception:

Atkinson, R.L., Atkinson, R.C. and Hilgard, E.R. (1983) *Introduction to Psychology*, 8th edn., Chapter 5. Harcourt, Brace and Jovanovich, New York.

For a detailed discussion of perceptual principles underlying the effective design of messages:

Fleming, M. and Levie, W.H. (1978) *Instructional Message Design*. Educational Technology Publications, Englewood Cliffs.

Communication

De Vito, J.A. (1994) *Human Communication: The Basic Course*, 6th edn. Harper Collins, New York.

McQuail, D. and Windahl, S. (1993) *Communication Models for the Study of Mass Communication*, 2nd edn. Longman, Harlow.

Learning

An introduction is given by:

Rothstein, P. A. (1990) *Educational Psychology*. McGraw Hill, New York.

Different approaches to learning are discussed in:

Bandura, A. (1986) *Social Foundation of Thought and Action*. Prentice Hall, Englewood Cliffs.

Bruner, J.S. (1966) *Towards a Theory of Instruction*. Harvard University, Cambridge.

Novak, J.D. and Govin, D.B. (1984) *Learning How to Learn*. Cambridge University Press, Cambridge.

Skinner, B.F. (1968) *The Technology of Teaching*. Appleton-Century-Crofts, New York.

Attitudes

Rajecki, D.W. (1990) *Attitudes*, 2nd edn. Sinauer Associates, Sunderland.

Decision-making

Beach, L.R. (1997) *The Psychology of Decision Making. People in Organizations.* Sage, Thousand Oaks.

Lindblom, C.F. (1990) *Inquiry and Change.* Yale University Press, New Haven. (Lindblom has made an important contribution to the empirical approach to decision making).

Ravnborg, H.M. (1996) *Agricultural Research and the Peasants: The Tanzanian*

Agricultural Knowledge and Information Systems. Centre for Development Research, Copenhagen.
Stevens, M. (1996) *How to be a Better Problem Solver.* Kogan Page, London.

Diffusion

The best known review of literature relating to diffusion is:
Rogers, E.M. (1995) *Diffusion of Innovations*, 4th edn. Free Press, New York.

Another important reference for other subjects in this whole chapter is:
Havelock, R. G. *et al.* (1969) *Planning for Innovation: A Comparative Study of the Literature on the Dissemination and Utilization of Knowledge.* Center for Research on the Utilization of Scientific Knowledge, Ann Arbor, Michigan.

The application of diffusion research is discussed in:
Lionberger, H. F. and Gwin, P. H. (1991) *Technology Transfer: A Textbook of Successful Research Extension Strategies Used to Develop Agriculture*, Publ. MX381. University of Missouri, Columbus, Missouri.

An interesting application in organization science is discussed in:
Attwell, P. (1996) Technology diffusion and organizational learning: The case of businees computing. In: *Organizational Learning* (M.D. Cohen and L.S. Sproul, eds) Sage, Thousand Oaks.

One of the origins of innovations is discussed in:
Okali, C., Sumberg, J. and Farrington, J. (1994) *Farmer Participatory Research.* Intermediate Technology Publications, London.

A review of the literature on the consequences of innovations, which showed that it is necessary to change some popular opinions, is:
Lipton, M. with R. Longhurst (1989) *New Seeds for Poor People.* Unwin Hyman, London.

6 Extension Methods

'Knowledge without practice makes but half an artist.'

18th century proverb

Ideas presented in the previous chapters should help extension agents to think systematically about their work. In this chapter we will discuss several of the methods extension agents commonly use to help farmers form opinions and make decisions. An extension agent's choice of any of the many methods available will depend on his or her specific goals and on the circumstances in which he or she works (1). The extension agent also must decide how to use these methods. The ideas presented in this chapter should help in making these decisions.

Mass media, group and individual or face-to-face extension methods are discussed in turn. Print and electronic media such as newspapers, radio and television help extension agents to reach large numbers of farmers simultaneously. However, there is little opportunity for these farmers to interact among themselves or to provide feedback to the extension agents. Group methods reach fewer farmers but offer more, if somewhat variable, opportunities for interaction and feedback. Formal lectures normally provide fewer opportunities than group discussions, although informal talks usually involve adequate interaction and feedback opportunities. Individual extension consists mainly of a dialogue between extension agent and farmer. We also discuss briefly how different methods can be combined in one programme, and how audio-visual aids can increase the effectiveness of these methods. We end this chapter with a brief discussion of how audio-visual aids, folk media and modern information technology can be used in communication with farmers.

6.1 Mass media

When discussing the use of mass media in extension, we must consider the role these media can play in an extension programme and how they can be used effectively. Hence we discuss which effects we can expect from use of mass media, and how these media can be used to ensure the meaning of our messages is as clear as possible. Finally, we pay very brief attention to the choice of mass media to be used, and to the differences between mass media and interpersonal communication.

6.1.1 Media effects

Newspapers, magazines, radio and television generally are the least expensive media for sending messages to large numbers of people. However, we should examine their effects before concluding that an efficient extension service should emphasize use of these media.

People have expressed different opinions about the extent to which mass media influence our thoughts and actions. In the 1940s it was thought that media could exercise great influence. They were credited with assisting Hitler to gain power in Germany, although there had been no systematic research at that time. Several reports were published in the 1950s which showed the media had limited influence in the American presidential elections. Joseph Klapper published his influential work *The Effects of Mass Communication* in 1960. Amongst other things, he analysed several selective processes which can cause this limited influence effect under certain circumstances (2). Views have changed again so that now it is felt that media may have more influence than Klapper and his followers had thought possible (3).

Our discussion of perception in Section 5.1 and the adoption and diffusion of innovations in Section 5.6 gave us some insight into how mass media work. We saw that perceptual principles can be used to attract attention, and that media are important for making people aware of innovations and for stimulating their interest. However, the media appear to have little direct influence when it is finally time to make a decision. At this critical stage we value the judgement of known and trusted people with whom we have discussed the issues. Research shows that the media can accelerate existing change processes, but they seldom bring about changes in behaviour by themselves. This is because the sender and receiver tend to employ several selective processes when using mass media, which often result in the receiver distorting the sender's messages. These processes include:

- selective publication
- selective attention
- selective perception
- selective remembering

- selective acceptance
- selective discussion.

Selective publication

Newspapers do not print and radio and TV stations do not broadcast all the information available from all sources. There is not enough space or time. Consciously or unconsciously, editors will decide what they will include or exclude. This selection process, sometimes referred to as '*gatekeeping*', is influenced by the personal values and beliefs of the editors, by the orders or wishes of the owners of the medium, of the advertisers or of the government. The selection process can seriously limit the public's access to information. Where there are strictly commercial objectives, the media will publish what they think the public wants to read, hear or see. In some situations editors or publishers select certain information because they want to spread an ideology or combat one-sided information presented by others.

Many countries value the freedom of the press highly. This freedom implies that the editors have the right to publish any information they wish, although certain restraints may be placed on ideas or news which could influence national security. In other countries the media, especially television and radio, are used by the government to broadcast or publish propaganda for their policies. These media tend to publish what the government believes is good for the people and the country, or sometimes is good for their leaders, an approach which makes the media rather dull and untrustworthy for their audience. Technical developments such as short wave radios, satellites and the Internet make it increasingly difficult for government to prevent information from outside their country reaching the population.

Selective attention

Nobody can read everything that is published, so we must select rigorously. Research shows we select on the basis of the use we expect to make from gratifying our needs for reading, listening or watching. This might be our need to solve certain problems or to know what is happening around us, or our need for relaxation or companionship. It also can be from the need to participate in discussions with our friends or to confirm our own point of view. The main focus of research has been on this latter need.

Selective perception

We tend to interpret messages we disagree with so that we make little or no change to our own opinions. Thus we may decode the message differently from the way it was encoded.

An extension campaign in a Middle Eastern country was designed to persuade farmers to tie their cattle halters in a different way that would not injure the cows' ears. The posters used in the campaign included drawings intended to show the correct and incorrect ways of tying the halters.

Unfortunately the designers did not test their messages carefully with farmers before printing the posters. Later research demonstrated that farmers' attention was drawn to the 'wrong' method marked with a red cross, and showed that they regarded the recommended method as foolish and impractical (4).

Selective remembering

Nobody can remember everything they have ever heard or read. People tend first to forget points which do not match their existing opinions (compare this with Bruner's theory of learning discussed in Section 5.3 'The cognitive map') or which place them in an unpleasant dilemma. We are more likely to remember details of discussions with friends or colleagues than something we read in a newspaper or heard on radio.

Selective acceptance

We may remember but not necessarily believe advertisements telling us that Brand A soap powder is more effective than any other brand. We tend to accept ideas more easily when they agree with our own opinions. We may offer less resistance to messages if we are unaware we are being influenced. Thus, we are less likely to reject cleverly devised television 'commercials' in which actors portray behaviour desired by the advertiser than we are to reject direct appeals for us to change behaviour in the same way. Similarly, it is possible to 'act out' key ideas in an extension campaign in a traditional society using travelling theatres or shadow puppets. We tend to select the knowledge we accept, but often we pick up values from movies and plays which may not be taught consciously. Certainly, we may wonder which of these changes in people's values are desirable.

Selective discussion

We do not have time to talk to other people about everything we read or hear about in the media. Research shows that we talk mainly about things which agree with our friends' opinions or which we think might interest them. Similarly, we tend to talk primarily with friends who have similar opinions to our own, thus confirming these opinions. In professional circles scientists tend to talk to fellow scientists from the same or closely allied disciplines, and often mainly with those who share their views on major issues. Agronomists who favour organic agriculture talk mainly to agronomists who have the same views.

Analysis of these selective processes is related to the individual-blame hypothesis discussed in Section 5.6.2. We also see here an increasing interest in the system-blame hypothesis by analysing people's access to the mass media.

Access

People in less industrialized countries have limited access to mass media for several reasons. Many people, especially women, are illiterate. This problem is overcome in some families with literate children who read publications regularly to their parents. Newspapers often are not distributed in rural areas, and, even when available, may be too expensive for most families. Television also is too expensive for many, or there may be no electric power supplies in rural villages. Where it is available, there are often many blackouts or electric current fluctuations which damage the sets.

Radio is available to most rural people around the world, especially since small transistorized sets were developed in the 1960s. However, these are not easily repaired by villagers and batteries sometimes are considered to be too expensive.

The physical availability of media is not the only critical factor. We must also consider the extent to which their messages are written and programmed for rural audiences. Most newspapers and television stations are located in cities and direct their information to urban audiences. They seldom discuss problems which interest a less educated rural audience.

There are many languages and dialects in some countries. For example, nearly every valley speaks its own dialect in the Highlands of Papua New Guinea. Educated urban people and less educated rural people nominally may speak the same language, but in such a different way that language used on radio and television is difficult for many rural people to understand. Furthermore, the speed of presentation on electronic media may be too high for language comprehension or for decoding pictures. There are clearly as many differences among rural people themselves as there are between urban and rural people. There may be well educated and culturally highly urbanized people living in the same village among poor, illiterate women who have little exposure to urban culture.

Recent studies suggest the mass media can play a greater role in the process of change than earlier had been thought possible. We do not challenge the existence of the selective processes outlined above. However, we must draw attention to the fact that media fulfil certain functions in our societies and in changing these societies. These include:

- setting the agenda of important discussion topics;
- transferring knowledge;
- forming and changing opinions;
- changing behaviour.

Setting the agenda of important discussion topics

The media can have an important influence on what we think and talk about, even though they cannot decide what we must think (5). For example, the media draw attention to problems faced by the population during a famine and to measures taken by government to overcome these problems. They may discuss why there are food shortages, despite the fact that similar problems may have occurred for years. The grain storage facilities may be inadequate or infested with pests, or corrupt officials may be diverting emergency food supplies to enrich their own pockets. Unfortunately, the media tend to discuss and highlight problems such as these only in times of crisis.

Farm magazines and rural radio programmes can play an important role by stimulating farmers to discuss points with extension agents or opinion leaders.

Transferring knowledge

We learn only part of what we know about the world through our observations and direct experiences, or from hearing about other people's observations and experiences. We gain much of what we know about the world from the media, although the view we develop in this way does not always agree with reality. The media specialize in news. While 'man bites dog' would attract attention as news, 'dog bites man' would be regarded as too commonplace to warrant media attention. Extension agents sometimes will try to create news in order to win media attention and coverage. Public demonstrations for political causes such as street marches and blockades by agricultural machinery also serve this end.

Knowledge is more likely to be transferred successfully if it meets a need or fills a vacuum. New ideas diffused through the media are more acceptable if they link up with existing knowledge than when they attempt to modify this knowledge (Section 5.3). People who are convinced that they have correct information, which can be proven scientifically to be incorrect, often will require a more personal approach to change their views. Many farmers in southern Australia placed great faith in dolomite as a pasture fertilizer despite extensive scientific evidence that it was for all practical purposes an inert and insoluble form of phosphate rock. Extensive media publicity failed to change their opinions. It required extensive personal discussion and demonstrations by extension agents and other convinced farmers to change these views.

Some kinds of knowledge can be transferred though the media, whereas this is not possible for other kinds of knowledge and skills. For example, the basic idea of Integrated Pest Management that chemicals can destroy the ecological balance by killing the predators of dangerous insects, might be taught through television, but the skills required to recognize and count the different kinds of insects can only be taught in the field.

Forming and changing opinions

Mass media may play an important role in developing opinions when members of the public do not have strong views about particular issues. Less effort is required to follow someone else's opinions than to form one's own, especially when the topic has no special personal significance. Media also have important effects in changing opinions when the position they advocate differs only slightly from one's own opinions.

We are inclined to think more carefully about important issues, finding out what other people think and generally discussing points before we make up our own minds. Others may take the initiative in these discussions if they think the subject to be important enough. Husbands in traditional circumstances tend to influence the political convictions of their wives in this way. Fortunately farmers often will not accept knowledge and opinions transferred through the media, but use them creatively to change their cognitive map and form their own opinions. They will also use other information sources in this process, as we explained in our discussion of Agricultural Knowledge and Information Systems in Section 2.4.

Opinions expressed in the media are not always unanimous, which makes it very difficult for members of the public to make up their own minds. People in this situation are inclined not to change opinions without realizing that the decision not to change also is a decision. Some medical authorities have proposed a relationship between use of herbicides such as 2-4D and 2-4-5T and the incidence of genetic disorders among newly born children. Other expert authorities disagree with this view. Farmers who make extensive use of these herbicides thus are placed in a quandary about which opinion to believe. Environmental groups often add their support to one side of the argument by making emotional pleas to ban all agricultural chemicals.

Changing behaviour

Mass media may be used to change patterns of behaviour, especially where these changes are small and relatively unimportant or where they help us to fulfil an existing wish. Advertising is very successful in this way. Cleverly worded advertisements draw consumers' attention to products they felt a need for but did not know existed, or they help people make small decisions about which brand of product to buy. For example, several companies may produce agricultural sprays with identical chemical composition but different brand names. Advertising influences farmers to choose one brand in preference to another.

There are other personal problems such as giving up smoking or losing weight where many people would like to change their habits but find it difficult to do so successfully. Important behavioural changes can follow when the media show people how to fulfil their wishes. It is still not clear under which conditions this will happen; nor are the effects of mass communication on interpersonal communication well understood. Mass media are often used for one way communication from an expert or a policy maker to farmers, but they can also be used in a more participatory way (6).

Farmers' information needs must be assessed before planning mass media messages. This process requires careful listening to farmers in order to integrate their felt needs and their needs as assessed by the experts. Such assessment requires a good understanding of the reasons why farmers farm as they do. For example, it is possible they do not realize that their present way of farming endangers the sustainability of their farm. Knowledge from research and from farmers' experiences can be integrated when presenting solutions to farmers' problems. It is usually highly appreciated by farmers if this is done in interviews with other successful farmers who do not have an exceptionally high level of resources. Information can also be presented which increases the farmers' skills in developing better solutions to their problems themselves.

Mass media have also been used successfully to inform policy makers and researchers about farmers' current situations and about the problems they face. This can decrease the urban bias in policy making found in many less industrialized countries.

Thirdly, mass media are often used by farmers' organizations to share experiences among farmers and to increase their motivation to work together to solve their problems.

The impact of mass media on rural populations

Nearly all the research on which this section is based has been conducted in industrialized countries. We cannot be sure yet how far the findings can be applied to rural people in less industrialized countries as there are many conflicting views on this subject. Some say that people in industrialized countries are saturated with information, whereas in less industrialized countries they are hungry for it. Hence the selective processes are not very important in these latter countries. The mass media play a very important role in introducing knowledge, opinions and entertainment from outside the local community. Furthermore, people in industrialized countries have learned from experience to distrust many messages they receive from the mass media, whereas in some less industrialized countries people are inclined to accept authority, including authoritarian messages from the media.

Others believe there is a wide gap in culture, language and interests between rural people in less industrialized countries and those writing or broadcasting for the media. Hence rural people will be disinclined to accept messages from these media, even if they had ready access to them.

We are not yet sure who is right. It seems probable that those media which have gained the confidence of rural people in less industrialized countries can have considerable impact. This seems to be the case with the radio programmes of the Acción Cultural Popular in Colombia. Some entertainment media also have had considerable impact. For example, it has been said that the juke-box has been more effective in wiping out the old Indian pagan songs than 400 years of preaching by Spanish priests. This is because the songs are replaced by music considered to be more attractive to modern young listeners. Uncritical acceptance of Western ideas therefore is a cause for concern, as is the uncritical rejection of all Western ideas and customs for that matter. An extension organization might consider using advertising techniques if there was one single message to get across to a large audience. There are several examples where this has been done successfully with nutrition education in less industrialized countries. For example, a one minute radio message repeated several times daily was made to teach mothers how, by using oral rehydration methods, they could save their child from dying of diarrhoea. Messages like this may reach mothers who are not motivated to attend nutrition classes. Simple agricultural messages also could be transmitted effectively in this way.

Choosing a communications strategy

Investigations of the effects of mass communications generally have given us little insight into how a chosen communication strategy influences these effects.

We tend to assume receivers are resistant when media fail to achieve behavioural change, although logically the source of the message also may be responsible. Source credibility, reliability and relevance are very important. Credibility generally will be high if the source is considered to be an expert, if the source agrees with the receiver on important points and if the receiver is convinced the source is trying to serve the receiver's interests as the receiver perceives them. Credibility is not necessarily universal – a source may be perceived as credible by one receiver and not by another, although well known and highly respected people tend to have high credibility with a wide range of receivers.

We have indicated in passing how we must choose a communication strategy which links up the knowledge, language use and attitudes of a target group with members' desire for information. People must want information before we can achieve behavioural change. This presents a problem in agricultural extension when farmers believe they manage their farms very efficiently and have little to learn from so-called 'experts'. They may accept that expert advice is useful, but they do not always accept readily that it may be useful to them. Expert advice always is meant for the other person. Extension agents faced with this situation first have to convince farmers they have something to gain from learning about and applying the information. Our discussion of decision-making showed that relevant information, especially when it is presented through demonstrations, can create awareness of a problem or of a new opportunity to achieve our goals. In turn this can create a need for information about how to solve the problem or how to achieve the goals.

Message senders may have to employ indirect strategies which appeal to other stronger interests of their receivers when the latter show little enthusiasm for information. They can build technical information into an amusement programme or combine their themes with others for which receivers feel a need.

Most empirical research into the effects of mass communication has been limited to a maximum time-scale of two years. Despite their obvious importance, few long term studies have been carried out because they are difficult and expensive to organize. Radio, and more recently television, probably have had a significant impact on the lives of many rural people. The advent of satellite technology and the widespread use of transistor radios and television sets powered from car batteries have extended news and information reception to previously remote villages and communities. Some writers have hypothesized that the electronic media are important tools for forging bonds of national unity and a sense of purpose in development programmes in less industrialised countries (7). Although these arguments intuitively make sense, it would be difficult to test and virtually impossible to reject them with research methods currently available.

6.1.2 Manner of presentation

'Your easy reading
is damned hard writing.' Hemingway

Technical publications produced by the extension services must present their information in a comprehensible form. But what is 'comprehensible'? German research has shown that there are four important factors in establishing comprehension (8). We believe that further research will demonstrate that these same four factors could be applied also to other languages. They are:

(1) *Use simple language*. Technical terms should be explained in short and simple sentences, using common words which have concrete meanings. Abstract language and 'jargon' should be avoided.
(2) *Structure and arrange arguments clearly*. Ideas should be presented in a logical order, clearly distinguishing between main and side issues. Presentation must be clear, with the central theme remaining visible so that the whole message can be reviewed easily. Careful use of layout and typography help to separate key points or sections of the message.
(3) *Make main points briefly* . Arguments should be restricted to the main issues and clearly directed towards achieving stated goals without unnecessary use of words.
(4) *Make writing stimulating to read.* The style should be interesting, inspiring, exciting, personal and sufficiently diversified to maintain the reader's interest.

These four factors appear to be independent of each other. A simple text may be structured or unstructured. It may be comprehensible if it is very simple and clearly structured. Texts may be too long or too short, although brevity appears to be the best strategy. A stimulating writing style may improve comprehension of a well-structured text but can make a badly structured text harder to understand.

Texts may be rewritten, taking into account these four factors for improving readers' comprehension, retention of information and satisfaction when reading. Research has shown that these principles apply to people with low intelligence and poor education as well as to well educated, intelligent people. Comprehension also is likely to be improved by tuning the text to the interests and knowledge levels of the target group. People may require special training to ensure they write simply and clearly.

The most effective manner of presentation in many countries also will depend on their tradition with 'oral literature'. Most farmers do not read much in countries with such a tradition, but they listen to story tellers, singers, religious plays, etc. Extension agents should study these traditions in order to be able to adjust their messages and manner of presentation to the expectations and experiences of their target group.

Advertising companies often pre-test their advertising messages to ensure they are clear and easy to understand. A small sample of the target group is tested and their reactions analysed to see whether people grasp the intended meaning. This feedback procedure gives the message producer an opportunity to change the content or message organization before expending time and money on major campaigns (9). It is important to pre-test the pictures as well as the words used in extension publications. Farmers who are not accustomed to reading and interpreting illustrations often misunderstand their meaning. They should be asked what they see in the picture as a whole as well as specific parts. Their answers then should be treated as valuable information to improve the extension message, and under no circumstances should they be ridiculed for gross misunderstandings. Unfortunately, in most countries extension messages are seldom tested in this way. We pointed out earlier in Section 5.2 how the meanings intended by scientific sources and those decoded by farmer receivers do not always coincide. This form of pre-testing is important for all authors, but especially if the author does not know the audience very well.

We should try first to answer the following questions before spreading a message via a mass medium:

(1) Precisely *which goal* am I trying to achieve?
Which problem am I trying to arouse interest in?
(2) Precisely *who* am I trying to reach?
What is their level of education?
Which problem are they interested in?
How much do they know about it?

Authors must be very familiar with the problems of their audience. They should ask themselves what is the best way to help solve these problems. This makes it extremely difficult to write for two different types of reader at the same time. For example, an agricultural research worker who attempts to write the same research report for fellow scientists and for farmers is unlikely to reach either group successfully. Messages must be written specifically for particular audiences, using language appropriate to their level of understanding, selecting messages which interest them, and using a medium they will receive and pay attention to. Farmers rarely read scientific journals because they often have difficulty understanding the technical terminology. Similarly, scientists are more likely to read scientific journals than popular magazines when they wish to keep up-to-date with current research in the discipline.

As a rule readers are especially interested in:

- subjects which have direct consequences or implications for them, their families and their friends;
- people they know or can identify with;
- events of topical interest;
- events which happened nearby;
- conflicts;
- successes or new developments;
- anything exceptional or out of the ordinary.

Editors will select articles for publication which they believe will attract most interest among their audience.

An author writing for the media may present relatively straightforward factual information or may dramatize the message. The dramatic approach is likely to have more emotional appeal and may reach a wider audience than the factual approach. Emotional appeals which stimulate discussion among their audience have a high probability of changing behaviour. H. Beecher Stowe's book *Uncle Tom's Cabin* probably had more influence on the abolition of slavery in the United States than any unemotional and purely factual presentation might have had. However, not all attempts at dramatization succeed, and we know of no empirical research into use of drama and the factors influencing its effects. It has been shown that dramatization can increase the appeal of the message, while at the same time it may decrease the audience's inclination to assess the message critically. Emotional appeals sometimes are used in agricultural extension messages to stress the dangers in using certain chemical sprays and farm equipment. Films can serve an important function by increasing farmers' emotional involvement in the topic.

6.1.3 Differences between mass media

Choice of the medium to be used is an important question for extension agents, although it is less important than the way they will use the medium.

This last question, which we discussed in the previous section, has more impact on the effect of the extension programme.

There appear to be three main patterns in the rapid development of mass communication techniques:

(1) *Increase in scale*. Media are being developed which can reach increasingly large numbers of people. Satellite technology has made it possible for hundreds of millions of people around the world to watch the same television programmes. Automated printing techniques using computers for preparing copy have extended readership potential for newspapers and magazines, but are economical only if circulation is high.
(2) *Decrease in scale*. New techniques and equipment suited to very small audiences are being developed at the same time as mass audiences are increasing. Video cassette recorders and players and copying equipment are now freely available in many countries. It is likely that extension services will distribute recorded video cassettes which farmers can play back through home or community television receivers. This development makes it possible to adjust media to the specific needs of a small audience.
(3) P*ersonalizing*. Computerized information services such as Viewdata, Videotex, World Wide Web and Internet, including CD-ROM technology, are spreading rapidly. These allow people to select specific technical and general information at will through their home television receiver or computer which in turn is connected to a central computer service through the telephone system. These services are discussed in section 6.6.

Extension agents in the twenty-first century no doubt will have further new communication techniques available. Although many current techniques will still apply, extension agents will have to learn to use the new techniques creatively. Thus, information specialists currently have to develop new skills in preparing messages for the small computer screen format. Unfortunately, research into possibilities for using new techniques lags far behind the research that lead to their introduction.

The ways in which people use different media have been studied frequently in industrialized countries. They spend more time with television and radio than with the printed word. The urban middle class in less industrialized countries also now spends considerable time watching television, but it is not yet a very important medium in rural areas of many of these countries.

Well educated people in industrialized countries pay more attention to print media, especially magazines, and less to television than poorly educated people. Print media have the important advantage that they give educated people the opportunity to select articles which interest them. Educated people also pay more attention to television news and information programmes and less to entertainment programmes than the poorly educated.

Thus we discern a tendency 'to those who have (information), to those will be given'. Attempts have been made to reach the poorly educated target group by 'sandwiching' an information programme between two popular entertainment programmes, for example. Comic books also have been used successfully for this purpose in some less industrialized countries.

Trade journals are very important for directing technical extension information to a selected target group. In this way they contrast sharply with media which provide news and entertainment. Strong bonds can develop between readers and their trade journal. Farm magazines usually are very important sources of professional information for farmers in industrialized countries. They range from general publications like *The Farmers' Weekly* to specialized magazines, for example, for broiler producers. Magazines like these also are beginning to play an important extension role for the better educated and newly literate farmers in some less industrialized countries. We expect the importance of farm magazines to grow quickly as literacy rates and the commercialization of agriculture continue to increase rapidly.

Extension newsletters produced by cheap offset printing techniques can be very useful, especially where the extension agents have access to good mailing lists and reliable addressing systems. Computerized word processors have simplified these systems by making it easy to update address changes without complicated mechanical changes.

Printed reference books have an increasingly important role to play in extension problem-solving strategies, especially in view of the rapid increase in the amount of knowledge available.

Radio is the most important mass medium for farmers in less industrialized countries. Farm radio programmes for agricultural development must be broadcast at times when farmers and their families can listen, usually early in the morning before going to their fields, or in the evenings after work. Broadcasters have to win their listeners' confidence by basing their programmes on local problems and by using language the farmers can understand. Interviews with successful small farmers usually are more effective than speeches by agricultural scientists.

We also must recognize there may be considerable differences within a specific medium such as magazines when we assess how different mass media work. General statements about radio may be relatively meaningless if we do not take into account the great differences which may exist between different radio stations and programmes. For example, in some countries there are national or government-sponsored services as well as commercial or privately owned radio stations.

In Table 6.1 we compare different media which can be used to reach farmers in industrialized countries. This table is based on the authors' impressions, and certainly is not valid for every specific situation. It is intended to stimulate the reader to think about characteristics of the media he can use. It is not yet possible to make a similar table for less industrialized

Table 6.1 Some characteristics of several mass media as extension tools in industrialized countries

	TV	Radio	Newspaper	Farm magazine	Leaflet
Receiver's message interpretation	xx	xxxx	xxx	xx	xx
Degree of feedback	x	x	x	xx	xx
Peer influence	xxx	x	x	xx	xx
Receiver's activity	x	xx	xx	xx	xxx
Audience size	xxxx	xxx	xxx	xx	xx
Audience education level	xxx[1] x[2]	xx[1] x[2]	xxx	xx	xx
Message cost	xxx	x	xx	xx	x
Cost paid by extension organization	xx	x	x	x	xxx
Degree of source credibility	xxx	x	x	xx	xxx
Access to medium	x	xx	x	xxx	xxxx
Extension agent's decision on message	x	x	x	xx	xxxx

The number of crosses indicates the extent to which the medium has this characteristic.
[1] in poor countries; [2] in rich countries.

countries because of large differences between these countries and the limited information available on how the media operate. Extension agents should observe which media farmers use, and discuss their use with them before deciding which to use for extension messages. For example, if an extension agent observed large numbers of cassette tapes being sold at Indonesian markets, or many comic books on display at Mexican markets, he or she should consider carefully whether to use these media also, first taking the precaution to ask whether or not his or her target group uses these media frequently.

6.1.4 Comparing mass media and interpersonal communication

Comparisons made between a specific medium and other media help us to develop insight into the advantages and disadvantages of that medium. Table 6.2, which more or less summarizes our discussion of mass media, compares mass media and interpersonal communication. In this table we have ignored differences between the various media, and between the different methods of interpersonal communication.

Table 6.2 Differences between mass media and interpersonal communication

Characteristic	Interpersonal channels	Mass media channels
Message flow	Tends to be two-way	Tends to be one-way
Communication context	Face-to-face	Interposed
Amount of feedback readily available	High	Low
Ability to overcome selective processes (selective exposure)	High	Low
Speed to large audiences	Relatively slow	Relatively rapid
Possibility to adjust message to audience	Large	Small
Cost per person reached	High	Low
Possibility for audience to ignore	Low	High
Same message to all receivers	No	Yes
Who gives information	Everybody	Experts or power holders
Possible effect	Attitude formation and change	Knowledge change

Adapted from Rogers, E.M. and Shoemaker, F.F. (1971) *Communication of Innovations*, p. 253. Free Press, New York.

☞ Discussion questions

1 What effects do you expect from a mass media campaign to promote the safe use of pesticides? What are the main factors influencing the effectiveness of such a campaign?
2 Judge a publication such as this section of this book using the four factors of comprehension discussed in Section 6.1.2.
3 Which new mass media have emerged in the last ten years? How can extension organizations make best use of these media?
4 Using mass media, how can we:

(a) stimulate interest among a poorly educated part of the population to participate in discussions about plans for developing the area in which they live?
(b) give this group sufficient insight into the plans so that they can participate effectively in discussion?

6.2 Group methods

'When you talk less
You can listen more.' Indian proverb

Group extension methods have an advantage over mass media because of better feedback which makes it possible to reduce some of the misunderstandings that may develop between an extension agent and a farmer. There also is greater interaction between the farmers themselves. This interaction provides the opportunity to exchange beneficial or useful experiences in order to integrate information from farmers and extension agents, as well as to exert influence on group members' behaviour and norms. Group methods differ considerably from each other in opportunities for feedback and for interaction.

Per capita costs of using group methods tend to be much higher than for use of mass media, especially if working with small groups. Therefore, we use these methods only when we need feedback for the extension agent or interaction between farmers to achieve our objectives. Otherwise the change can be achieved at lower cost with mass media.

Group methods often reach one select part of the target group because only those farmers who are more interested in extension and/or those farmers who are members of certain farmers' organizations come to the meetings.

Lectures, demonstrations and group discussions are the group methods considered in this section. We pay most attention to group discussions as the technique which offers the greatest possibilities for using the specific advantages of group methods.

6.2.1 Speeches or talks

Speeches or talks are an important means of transferring information in extension. Despite their relatively higher cost per capita in comparison with mass media, speeches or talks have some specific advantages:

(1) The speaker can modify the content of the talk to meet the specific needs and interests of the audience as well as their level of education.
(2) The speaker can take account of audience response during the talk and modify his or her approach accordingly.
(3) The audience gets to know the speaker better and receives a clearer impression of his or her feelings about the topic through gestures and facial expressions. This also is possible with television to some extent.
(4) Talks usually provide an opportunity for the audience to ask questions and discuss issues in greater depth.

The disadvantage with talks is that the spoken word usually is forgotten more easily than the written word. It is difficult to maintain an audience's attention on the subject of a talk for much more than 15 minutes unless the speaker is extremely dynamic and interesting. People's thoughts tend to wander to other unrelated questions.

Publications can be re-read if something is not immediately clear, but a listener quickly may lose the thread or forget the main points in a talk. Most farmers have not been accustomed to gaining their information from formal lectures and talks. Hence they are unlikely to remember much detail from talks.

Speeches and talks also are poor methods for teaching how to apply information. They must be combined with discussions and practical demonstrations to achieve this objective.

Talks often are followed by a discussion period in which questions are asked, either to clarify some points or out of politeness to the speaker. A lecture followed by a discussion may be used to focus public attention on a problem. For example, a speaker may give little information during the talk, but may ask many questions. These questions then become the topics for discussion following the talk. In the case of larger groups, it is desirable to divide the audience into small discussion groups so that all group members have the opportunity to present their views. Information discussed in the small groups then may be summarized in a plenary session in which the whole audience participates.

A review of investigations by Bligh (10) showed that lectures are as effective as group discussions in transferring knowledge, but much less effective in stimulating attention and changing attitudes.

Method

A good speech or talk must be well prepared. Some important points to consider are:

(1) The speaker must study the literature carefully, remembering that the art of giving a good talk is not to tell everything he or she knows, but to highlight the main points, which may be lost if there is too much detail.

(2) The speaker must take account of the audience, what problems interest them, how much they know already, and so on. It is always best to relate subject matter of the talk to interests, experiences and needs of the audience. For this reason, time spent researching these points will be well spent, especially if the speaker can incorporate local examples into the talk.

(3) The speaker must decide how to arrange the content of the talk. Unfortunately there is little empirical research evidence with which to plan the structure of a good talk. Some have argued that a logical thought pattern is much easier to remember than a more rambling approach. Certainly, it is easier to follow the trend of a talk when the speaker outlines the main points first. It also helps if these main points are illustrated with different examples, and if a clear summary is given at the end of the talk. There are others, however, who argue that people seldom remember all they hear. The main objective in their view is to stimulate the audience to think about the topic.

Most experts on public speaking agree it is undesirable for speakers to read their texts. Radio experience has shown that speakers who have some difficulty in formulating their thoughts and who have a slightly hesitant manner are very effective in transferring their ideas. They stimulate their audience to think for themselves. However, inexperienced speakers are advised to write out their talks first to check they can deliver them in the time available. They then should use notes which outline key points to jog their memory rather than the full text. A summary of the main points distributed among the audience assists them to follow the talk and helps stimulate discussion afterwards.

Information is more likely to be remembered and acted upon in the future if the audience not only listens passively but also thinks actively about the issues being discussed. Hence it is a good practice to combine talks with other extension techniques. For example, the audience may meet in discussion groups before the talk to consider key points. The speaker then can enlarge on questions arising in the discussion groups or can make it clear why he or she agrees or disagrees with conclusions reached in these groups. This technique often increases the audience's interest in the talk. On the other hand, group members may try to defend their original points of view, which means the speaker must be sufficiently expert and versatile to cope with points he or she may not have planned to include in the talk.

Another technique is to divide the audience into groups after the talk and provide them with specific questions related both to the content of the talk and to problems of the audience. This method may have less influence in motivating the audience to learn than the previous procedure. However, it can stimulate people to assimilate newly acquired knowledge and to discover ways of applying it in their own situation. People often are not inclined to disagree publicly with a speaker who has higher social status than the audience, especially in countries with an hierarchical social structure.

In Europe and the United States extension agents now tend to involve audiences more in the communication process than they did 20 years ago. This has not happened yet in many less industrialized countries.

6.2.2 Demonstrations

Demonstrations may stimulate farmers to try out innovations themselves, or even may replace a test of the innovation by the farmer. They can show causes of problems and possible solutions without complicated technical details, especially where the demonstration is of *results* of certain actions. For example, farmers can be shown the results of applying fertilizers at different times to overcome soil deficiencies without them having to understand the biological processes behind these effects.

We also have *methods* demonstrations where a technique is shown to people who are convinced already they want to use it. For example, fruit growers can be shown how to prune a tree. They have to return to the site later to see any longer term effect such as the increased yield caused by pruning.

Action demonstrations are a third type which try to show that large sections of the population desire changes in government policy or in their society. We will not discuss these demonstrations, because they are seldom organized by extension agents.

Results demonstrations can be given on demonstration fields or on demonstration farms. Demonstration fields can be used to compare results of traditional practices with a new practice or package of practices. For example, they can show that farmers can increase their yields by applying the right levels of fertilizer or by using high yielding varieties in combination with careful land preparation, fertilizers and pest control.

Demonstration farms can show farmers the effects of a change in their farming system, such as a change in crop rotation or the introduction of dairy cows which require feed to be grown. Farmers often will accept that the experiences on the demonstration farm are valid for their conditions only if the demonstration farm and farmer are similar to their own situation.

A methods demonstration can show a number of different brands of the same machine performing the same task in the same field, such as harvesting. This provides farmers with valuable information to choose the best brand for their situation by showing the performance of each brand under the same conditions. Extension agents can enhance the learning process by measuring machine performance, such as grain losses in threshing machines. They can discuss the results of their measurements and explain to farmers the different criteria they have used in judging machine performance. One problem is that machine performance often depends as much on the skill of the operator as on the quality of the machine. For example, if an unskilled operator runs a perfectly sound threshing machine too fast it will break many of the grains.

Result demonstrations are very important for making people aware of innovations in countries where the mass media play a limited role because of illiteracy or limited access to media outlets. Unfortunately there are few studies which explain the factors that influence the effects of demonstrations.

Methods demonstrations also can be used to teach farmers how to perform certain tasks such as spraying pesticides. When planning the demonstration the extension agent should analyse carefully the tasks a farmer has to perform, and the training objectives must be specified. Demonstrations of this type can only be given to small groups of farmers. Otherwise they will not be able to see exactly how to perform the task. Also, they should be given the opportunity to practise the task themselves and receive feedback from the extension agent on their performance. Tennis cannot be learned solely by watching how the trainer plays it – the student must practise the game. Many agricultural operations such as setting a plough or milking a cow also require practice.

A great advantage of demonstrations is seeing how a new method works in practice. There does not have to be such a high level of trust between farmers and extension agents when the farmers can see things for themselves. Nor does the extension agent have to be so concerned with the problem of encoding messages in words which could be decoded in different ways. Demonstrations are very useful for convincing people who have not learned to think abstractly. However, it must be possible to present the central idea behind a demonstration in a visual form, and the relationship between cause and effect cannot be visualized easily. For example, the effect of an innovation on a farmer's income cannot be demonstrated, although it might be inferred from a result demonstration.

Our expectations, needs and ways of thinking influence how we interpret what we observe. For this reason demonstrations may be interpreted in different ways, thus making it difficult for an extension agent to get the message across accurately and clearly.

Psychologically difficult situations may arise with demonstrations, especially if they are called 'examples'. These imply that you would have obtained better results if you had carried out your own work as well as shown in the example. Understandably, this creates resistance in which people look for a simple explanation. Farmers may claim that the demonstration differs from their situation because conditions are 'different'. For example, the cost of fertilizer used in the demonstration has been subsidized, or the extra labour required is provided free by a government-supported research station. There will be much less resistance if the demonstrators are selected by the community and asked to test whether the innovation works well in their village.

Demonstration farms may represent an intervention in the normal social structure of a village where the best farmers have the highest social status. For example, it would be a big mistake in this type of village to select a lower status farmer because he or she is well educated and co-operates enthusiastically with the extension service. Such a step could be perceived as an attempt to change the existing social structure, and hence would be resisted strongly. This also raises the question of the wisdom of having only one demonstration plot in a village. It may be good practice always to have at least two demonstrations in different parts of the village or on farms of different sizes.

Demonstrations must show clear differences between traditional and recommended practices, and they must be well managed. Demonstration plots should be kept simple, preferably comparing only the traditional with the improved method on a good-sized field.

It is common for demonstrations to fail when extension agents consider it beneath their dignity to carry out manual labour, or when supplies of fertilizers and other necessary inputs are not well organized. Failures lead to loss of faith in the innovation which may take a long time to overcome because of the psychological resistance to demonstrations.

Extension agents should have a clear idea of what they are trying to demonstrate, and what is the role of the demonstration in the extension programme. Agents should know whether farmers really feel a need for the information they can gain from the demonstration.

Demonstrations must be well integrated into the extension programme if they are to be effective. Other methods then can be used to stimulate farmers to visit the demonstrations and to decide how they will use the new information on their own farms. For example, it is very useful to organize a field day at a result demonstration site during harvest where farmers can observe the results obtained and discuss their implications. It is also wise to give the farmers a leaflet which describes the package of practices so that they may avoid making mistakes when applying them.

This section on demonstrations is brief in relation to the importance demonstrations have in the extension programmes of less industrialized countries. This is because there has been little research on factors influencing the effectiveness of this method. Hence we have limited possibilities for suggesting how the reader should use demonstrations effectively in his or her extension programme.

6.2.3 Group discussions

The role of group discussions

Group discussions are used for many different purposes in different societies. Committee meetings are used at a political level to achieve consensus or unanimous decisions. They also are used in agricultural extension, as well as in community development and adult education programmes, to help members identify and find solutions to their problems.

They are very important because they provide opportunities to influence participants' behaviour. Extension agents play different roles in groups from those they play in speeches or talks. They are usually the expert source of information, and hence perceived to be above their audience in status. However, in discussion groups they participate as group members who join all other members in solving problems.

Group discussions, usually in groups of five to twenty members, also can play the following roles in an extension programme.

Increasing knowledge

Group discussions assist the process of transferring knowledge from an expert to the group, although the printed word and audio-visual material, as well as speeches, are cheaper, more structured and usually more effective. However, group discussion helps people assimilate knowledge by giving the audience an opportunity to ask questions, to relate the new information to what they know already, and, if necessary, to revise their views if they feel the expert is correct. Then if lingering doubts remain, they can discuss them with the extension agent.

We have pointed out already that the 'expert' leading group discussion does not necessarily possess the most useful knowledge. Other group members may have valuable information to contribute. A farmer concerned about the deeper biological aspects of diseases may have to consult an expert, but fellow discussion group members may be of more use in helping the farmer to decide on suitable management strategies. Farmers also have valuable indigenous knowledge based on their experiences and those of their ancestors (11).

Attitude change

Group discussions can fulfil several functions in the process of attitude change:

(1) *Creating awareness of problems and feelings*. We are inclined to evade problems if forced to choose between two alternatives, both of which have unpleasant consequences. Furthermore, often we are only partially aware of our feelings. Group discussion can help us eliminate these constraints on good decision-making by making us more aware of the problems and of our feelings. We do this by analysing our present situation, by considering changes that have occurred in the past year and those likely to occur in the coming year, and by working out the implications of these changes for ourselves. We can distinguish between feelings about ourselves such as self-esteem, about other group members, the extension agents and the problems under discussion, all of which can influence each other.

A group atmosphere of mutual trust helps us to see and face up to problems we have tried to evade in the past. For example, farmers may be able to see clearly that they face financial ruin unless they change the nature of their enterprise. It is sometimes easier for us to acknowledge our own feelings in a group where other members openly discuss their own feelings. This technique often is used in counselling programmes involving marital problems, and more recently in the feminist movement. It has been used to a limited extent by specially trained extension agents in marginal Australian dairying districts where some farmers have to be helped to face the reality that they should leave the industry. It is much easier to help them solve the problem once they have faced their problem openly and realistically and see that they are not the only one who has the problem.

(2) *Concrete formulation of problems*. The more clearly a problem can be defined, the more likely it is a solution will be found. Group discussions can help us specify our problems clearly and in more detail. A group discussion between farmers and extension agents can help them come to an agreement about the correct definition of the problem, which is the first step towards solving it. Many Australian dairy farmers in the 1970s, faced with the prospect of insolvency because of drastically reduced incomes, took the drastic step of seeking part-time jobs away from their farms, relying on their wives

and families to look after their cows. Others left the industry and sought new jobs in the cities. Previously they had seen their situation as a farm management problem, but after formal and informal group discussions they were able to make decisions based on a wider definition of their problem.

(3) *Change in norms.* In Section 5.6.4 we have shown the important influence norms have on our behaviour. These norms do not change if an outsider like an extension agent says they are old fashioned. However, they do change if the group itself concludes that they should change. Group discussion can lead to a change in norms if various participants already have accepted the change, but incorrectly believe that their fellow group members still adhere to a traditional viewpoint. This occurs sometimes in agriculture when individual farmers accept the need for government intervention in their industry, but publicly oppose it until they hold group meetings and pool their common viewpoints. Discussions in Indian farm radio forum groups helped to change attitudes towards rat control. Farmers were aware that rats caused substantial damage, but they thought most other villagers adhered to the Hindu belief that one should not kill an animal. When they discovered from the radio programmes and in group discussions which followed these programmes that this perception was incorrect, they changed their own attitudes and behaviour towards rat control.

(4) *Formation of opinions.* Group discussion can help participants form an opinion about a specific issue or new development. Formation of 'sound' opinions results from mutual testing of ideas among group members. This does not mean necessarily that opinions are uniform but it does ensure they have been considered more carefully.

Behaviour change

Group discussions may fulfil the following functions in behaviour change:

(1) *Individual decision-making.* Lewin demonstrated the important role of group discussion in a series of famous experiments (12). He compared the effects of group discussion and individual advice when teaching mothers to give their babies cod liver oil. One group of mothers about to leave hospital held group discussions about the best ways to feed their new babies, and took a group decision that each should give them cod liver oil. Another group received individual advice on the same topic. Four weeks later 85 per cent of group discussion participants had given cod liver oil, compared with 55 per cent of those given individual advice. Lewin sees the change of group norms as an important link in the change process.

The individual decision in the Lewin experiment was taken in the group. More radical decisions will be taken later, often after consultation with family members. Making a radical change requires more time (Section 5.6.4), and because individual situations differ, not all group members can take the same decision. Furthermore, people like to retain some degree of privacy, especially with respect to their financial situation which they may not wish to discuss in public.

Group discussions also can increase the need for individual counselling. Hence, to avoid disappointment it is advisable not to start group discussions if the extension agent is unable to satisfy this need. Participation in group discussions often can help participants to remove mental blocks that make it difficult for them to find satisfactory solutions to their problems. However, much of this decision-making takes place outside the group.

(2) *Collective decision-making*. In many committees group discussion leads to a collective decision. Group discussion can have an important extension role in helping people become aware of their collective interests and in deciding how they can best protect these interests. For example, soil erosion often starts because of bad cultivation practices on farms at the head of a valley. Farmers in the lower parts of the valley are forced to suffer the consequences of their neighbours' careless land-use. Discussion groups are a valuable tool for mobilizing community action to overcome the problem. The Landcare movement in Australia has achieved many successes in changing community attitudes towards revegetation, salinity and erosion control in this way (13).

(3) *Confirmation of the choice*. Doubts about whether or not we have made the right choice between alternatives may linger, especially when the choice is important and we have hesitated to make a decision. Furthermore, there may be some adjustment difficulties when farmers make an important change such as in the crops they grow. Farmer may feel a strong need after making their choice to discuss with others who have made similar decisions whether they have made the right choice and to look for solutions to the adjustment difficulties. These discussions also can strengthen their decision to implement the choice already made.

We can gain some insight into the function of group discussion by comparing the advantages and disadvantages of discussions and lectures. This is summarized in Table 6.3.

Discussion group methods

Rules for leading discussion often appear so obvious that they are frequently ignored in practice. They may acquire great importance for those who helped formulate them if the group itself draws them up on the basis of their own experiences of what is useful and what is not.

Directive or non-directive extension methods can be applied in group discussions (14). Under directive conditions extension agents describe the problem for group members as they see it, and give a solution. Under non-directive conditions they help members formulate and analyse the problem themselves, and help them work out a solution. They may contribute some

Table 6.3 Advantages and disadvantages of group discussion compared with speeches

	Advantages	Disadvantages
1	Participants discuss more aspects than the extension agent does on his or her own.	The transfer of information takes more time.
2	The participants are better judges than the extension agent whether possible solutions are practical.	Problems are discussed less systematically than in a speech.
3	In group discussions there is a strong tie with daily practice that is not usually present in a speech.	There is a danger that some participants will 'ride their hobby horses' or dominate discussion.
4	Language used in discussion is more familiar to participants.	A good discussion assumes participants have at least minimal knowledge. Otherwise discussion becomes pointless.
5	Participants can ask questions, present opposing ideas, and this may improve assimilation of what is said.	There is a chance that incorrect information given by one group member will not be corrected.
6	Group discussions stimulate participants' own activities much more than speeches.	Group discussions require a capable extension agent who can handle unexpected problems that may arise.
7	Participants have more opportunity to discover unknown aspects of the problem. This increases the probability that participants will adopt solutions to problems discussed by the group.	The socio-emotional climate has a great influence on the effects of group discussions. It is not always easy to influence this climate in a positive way.
8	Participants generally will be more interested, because they can exert influence on the choice of problems to be discussed.	Group discussions require a certain degree of homogeneity in the group.
9	Group discussions can have a significant effect on decision-making as well as on transfer of information.	Discussions should be held in groups of no more than 15 participants, whereas speeches can be held for much larger groups.
10	Group norms can be considered at a group discussion, and changed by the group, if necessary.	
11	Discussion leaders learn more about knowledge levels and problems of group members than speakers.	

information to the discussion, but only when it is clear the group needs it and other group members cannot provide it. The non-directive approach generally is to be preferred, although there are times when extension should be directive. There is little reason to use group discussion in the latter situation.

The difference in function of directive and non-directive discussion methods can be expressed in another way. The directive method is appropriate if the only goal is to find the best solution to a problem, so long as the extension agent is expert, is trusted by the group and has sufficient information from other group members to pose the problem correctly. On the other hand, the non-directive method is more appropriate where emotional considerations may influence the search for an acceptable solution. The group may decide to choose a different solution from the one the extension agent might have chosen in this situation.

The farmers remain dependent on the extension agents in a directive approach. They will have to seek the extension agents' help each time they encounter a similar problem. This is not serious for a problem which arises only once, such as construction of a new farm building. However, where the problem recurs regularly it is advisable to teach the farmers to make their own choices by giving non-directive extension information.

Extension agents with a sound technical background may identify farmers' problems and be able to suggest solutions easily. They may be inclined to offer this information quickly so as to save time and to satisfy farmers' and their own expectations of their knowledgeability and efficiency. Furthermore, they improve their self-respect and rise in the farmers' esteem as experts. Thus, there are rational, social and psychological reasons for an extension agent to give information quickly, as, for example, when a group member asks a question which gives the agent an opportunity to provide this information. The disadvantage of this is that the farmers may not have worked out the situation for themselves and may not realize they will need the extension agent's information to solve their problem. Thus there is little they can do with the information, and they may have forgotten it by the time they realize it is needed.

Our argument above does not mean an extension agent always must use non-directive techniques automatically for leading group discussion. It may be desirable to work directively in some stages of a group process, such as when the goal can be reached quickly or because the group cannot yet cope with a non-directive approach. We must remember also that there are many intermediate positions between the extremes of directive and non-directive leadership which may be appropriate for a particular discussion.

Some extension services make little use of group discussions because of their tendency to degenerate into idle coffee shop talk. These bad experiences were more likely to have been the result of extension agents' lack of training in discussion group methods. We believe this is a very important type of training because well led discussions give the opportunity to penetrate deeply into farmers' real problems, even those they may find difficult to recognize themselves.

6.2.4 Section summary

Discussion groups make direct interaction possible between an extension agent and farmers, and between the farmers themselves. This means that extension agents' information can be adapted more closely to farmers' needs and knowledge levels than when using the mass media. Interaction between farmers may be very important when deciding if an innovation should be applied, and if group norms should be changed.

Speeches and talks have much in common with the mass media. They are an important method for transferring information, while group discussion can play an important role in farmers' opinion formation and decision-making. Discussions may help farmers to become aware of the way in which their feelings influence their decision-making. They also help group members to draw on each other's experiences for problem solving, and may have considerable influence on participants' behaviour. The discussion leader's task on the one hand is to structure discussion so that good solutions are found to problems; on the other hand he or she must promote a socio-emotional climate in which participants feel they can express their feelings safely.

These group methods are particularly important if used in conjunction with other methods in an extension programme. Demonstrations and excursions have the advantage that farmers can see with their own eyes how methods work, and may be able to see advantages and disadvantages of an innovation. Farmers are more likely to change their behaviour in the recommended direction if they can discuss what they have observed with the person organizing the demonstration, with other members of their group, with the extension agent and with other farmers.

☞ Discussion questions

1 You have been asked to give a talk to a farmers' group. How do you prepare for such a talk? You can choose another audience that you do not know well, or a topic where you are likely to have a very different opinion from the audience.

2 Suppose you are well trained in soil erosion control methods and you are invited to be a discussion leader for a farmers' group in an area with serious erosion problems. In this discussion, which information do you think you should put forward and which should come from the farmers? What should be your role in evaluating different solutions to the erosion problem?

3 Think about a discussion group in which you have participated. What function did this group fulfil for you? Could this group have fulfilled one or more functions? If so, how?

6.3 Individual extension

Mutual discussion, sometimes referred to as dialogue or one-on-one discussion, is the most important method for individual extension, and hence we will restrict our discussion to this. We will outline the functions of mutual discussion in extension, and specify its advantages, disadvantages and the conditions for using it. We then classify discussion situations and outline several discussion models which can be used in these different situations. Finally, we point out some factors which influence the relationship between an extension agent and a farmer.

6.3.1 The role of mutual discussion

Farmers strongly favour mutual discussion as an extension method, and many extension services devote the greatest proportion of staff time to it.

Advantages

We see four main advantages in this method:

(1) It is a very good way of supplying information required for solving a unique problem such as a major investment decision. The unique nature of the problem often will go together with the unique nature of the person who has to take the decision, his feelings, capacities, situation, etc. Mutual discussion gives an extension agent the opportunity to get to know this person very well. The extension agent can observe crops, animals and the farm in general especially during farm and home visits, thus getting first hand information on problems and their possible causes.

(2) It is possible to integrate information from the farmer (for example, goals and means) with information from the extension agent (for example, causes of problems and research findings about possible solutions).

(3) The extension agent can help the farmer to clarify feelings and to choose between conflicting goals.

(4) The extension agent can increase the farmer's trust in him or her by showing interest in the farmer as a person, his or her situation and ideas.

Disadvantages

Naturally there are some disadvantages:

(1) Costs are high in terms of staff time and travel.
(2) An extension agent who works mainly through mutual discussion and farm visits usually reaches only a small proportion of the target group. In countries with a good transportation system an agent cannot visit more than 200 farmers regularly.
(3) Extension agents can give incorrect information, for example, if they are not prepared to admit that they do not have the necessary information or if they follow their personal whims. It is very difficult for their supervisors to control or correct this problem.
(4) The method is based on a high level of trust between farmers and extension agent. The farmers will neither seek help nor disclose confidential information about themselves if this trust is lacking, because they will be concerned that the extension agent might not use such information in their best interests.
(5) Mutual discussion often will be initiated by a farmer who feels he or she has a problem, usually when the problem is well advanced and has caused some difficulties. It might have been possible to find a better solution to the problem if the extension agent had been approached sooner. Extension agents sometimes initiate discussion after observing a problem during a farm visit. However, it is usually physically impossible to visit more than a small proportion of farms in a district to make such observations.
(6) Mutual discussion may help with a farmer's specific problem. Sometimes farmers do not realize that they have the same problem as their neighbours, and therefore they do not seek their assistance to solve the problem together. This method of extension seldom is a solution for the promotion of collective interests such as in starting a cooperative.

The first four disadvantages are less of a problem for extension contact by telephone. However, telephone contact limits opportunities for conveying feelings.

Requirements of the extension agent

Extension requires a high standard of performance by an extension agent and a positive attitude towards farmers, especially in the personal relationship of mutual discussion. Hence this method is most suited to situations where problems have socio-emotional as well as technical aspects.

Brammer states the following requirements for extension agents who wish to help farmers effectively (15):

(1) They must be aware of their own values. They must ask themselves 'Who am I?' and 'What is important to me?' in order to develop a clear picture for themselves of their goals. This also prevents them from forcing their own values on farmers. (See Chapter 4.)
(2) They must be aware of their own feelings, especially in their relationship

with farmers. Self-respect can help them overcome feelings of disappointment which are inevitable in such a relationship. Such disappointment stems from the fact that there often are no satisfactory solutions to some problems of emotional importance to farmers. For example, it is very difficult to advise a small farmer that the best solution to his or her financial difficulties would be to sell the farm and take an urban job. Some extension agents choose their career because of their need for trustful relationships with their fellow beings, but we question whether this is always favourable for farmers' personal development.

(3) The extension agents serve as an example to farmers in the way they solve problems and in their emotional behaviour, whether they want to or not. They are expected to apply the consequences of their theories and ideals themselves. If they have their own farm, farmers will watch with particular interest to see what they do and how they do it.

(4) Effective extension agents have a strong interest in their fellow beings and in social change, their interest arising from an altruistic desire to help people. They also may wish to help change the structure of society as well as help individuals.

Extension agents must fulfil all of these requirements to achieve productive mutual discussion. That is to say, they must conduct discussion so that they help farmers' personal development as well as helping them to solve their problems. They must also be prepared to give help, and have time available to invest an adequate level of effort. This is unlikely to be possible where they have to serve several thousand farmers.

Requirements of the farmer

Farmers also must be prepared to invest time in mutual discussion, although sometimes they are too busy or, if part-time farmers, not available during the normal working hours of government officers.

Adapting mutual discussion to different situations

The nature of mutual discussion between extension agent and farmer depends on whether the farmer already is aware of the type of problem he or she faces, and the extent to which information required to solve the problem is available from the extension agent, the farmer or from both. (See Section 2.3.) Solutions to problems connected with feelings, value judgements and the investments which can be made, require information from the farmer. We pointed out earlier (in Section 5.5), that the farmer often is only partly aware of his or her own feelings and has not always chosen clearly between conflicting goals and value judgements. The extension agent can try to help the farmer to become aware and to choose.

The extension agent and the farmer often will each have part of the information required to solve a particular problem. At other times only the extension agent will have this information. These mutual discussion situations are reviewed further in Table 6.4. Let us demonstrate how the function of the extension agent differs by taking an example from each of these different situations, remembering they are only points of a continuum:

Table 6.4 Situations for using technical advice and counselling as methods of discussion

		Farmer's awareness of problem	
		Aware	Not aware
Information available to:			
Extension agent	(Technical advice)	1	4
Both		2	5
Farmer		3	6 (Counselling)

(1) A farmer sees a disease in his or her crop which he or she does not recognize. The farmer asks the extension agent what it is and what to do about it, as the farmer is confident the extension agent will have the correct answer. Only information can solve this problem.

(2) A farmer decides his or her farm is too small and asks the local extension agent what investments to make to enlarge it. The extension agent can draw up several alternative budgets, but first must ask the farmer for information about his or her financial position and interest in different enterprises. The farmer is less likely to implement a specific solution worked out by the extension agent than to use one they have worked out together.

(3) A part-time farmer, who is growing vegetables on a small segment of land, asks the extension agent whether or not to leave his or her job to concentrate on growing more vegetables. This decision requires information about the farmer's urban job income, likes and dislikes about this job, the income the farmer makes now from the vegetables, predictions of future vegetable prices, risks involved in the job and in vegetable growing, the farmer's reaction to these risks, etc. The farmer has most of this information, but may need help to use it systematically in decision-making.

(4) The farmer does not know that his or her only cow may catch a disease because of the way he or she is feeding her.

(5) A farmer reports to the income tax office that he or she makes $5000 a year, and thinks that he or she is doing rather well. However, if we make deductions from income for rent of land, interest for other investments and wages for the work of the family on the farm, it would show a net loss of

$5000. It is doubtful either that the farmer would become aware of the problem if given this information or that the farmer would be prepared to solve it. Such information would threaten the farmer's feelings of self-respect to a considerable degree, so he or she would have difficulty in accepting and assimilating it. An extension agent can help the farmer understand and face the problem if they have a good confidential relationship. The extension agent must first ask the right questions, thus helping the farmer to assess his or her own situation. Then the agent can help the farmer solve the problem in the same way as (2) above. This example immediately demonstrates how difficult it is to start a discussion in such a situation. It becomes easier if the farmer is aware of the problem but lacks insight into its nature. The extension agent then can help the farmer gain insight through counselling.

(6) The manager of an extension organization complains to a consultant that his or her Village Extension Agents are lazy and ineffective, and asks how the situation can be improved. The consultant identifies ineffective management of the organization as the main problem. It is doubtful whether the manager would accept such information if the consultant were to tell him or her bluntly. Apart from the obvious threat to self-esteem, the manager had invited the consultant to look at the VEAs and not at management. Hence the consultant will have to create a situation through counselling in which the manager discovers the real problem him or herself – a very difficult task.

Mutual discussion models

Various models of discussion have been developed for the different types of discussion situations. The *diagnosis-prescription model* (as used by the medical profession) can be used if the extension agent and the farmer are both convinced that the extension agent can, will and must solve the problem. It must be accepted that the farmer becomes dependent on the extension agent in this way, which may not be a bad situation for technical problems such as how to cure a plant disease. However, the *counselling model* is preferable where the farmer's feelings and value judgements are involved, for example, in deciding at what age he or she should hand over the farm to a son or daughter. The extension agent can help the farmer explore such a problem in relation to his or her feelings and values, by weighing them up against each other.

The *participation model*, in which farmer and extension agent both contribute to the decision-making, will be the one most commonly used.

When choosing a discussion model the extension agent must take into account that the effect of mutual discussion depends not only on the quality of the solution found, but also on its acceptance by the farmer. A very good solution which is rejected by the farmer is of less value than a moderately good solution developed by the farmer him or herself and hence acceptable to him or her. Furthermore, the farmer working in association with the extension agent usually can find a better solution than the extension agent might find working alone, because each has part of the relevant information. It is a good idea to test the hypothesis that the farmer is wiser than the extension agent if the farmer does not follow the extension agent's advice, because the latter, with his or her specialized knowledge, may have overlooked certain relevant aspects of the problem.

It also helps the farmers' personal growth and development if they can solve their own problems. Some people who advocate dogmatically that farmers should make all decisions themselves regarding their own future forget that a farmer's personal growth and development may be impeded if it appears that incorrect decisions lead to a poor solution. For example, the farmer may not have used information available from recent research.

We now will consider these mutual discussion models in more detail. Our presentation of these models may give the impression that extension agents can persuade farmers to do exactly what they want by applying certain psychological techniques. This is incorrect. Such psychological techniques can help an extension agent who genuinely strives to help farmers achieve their goals, providing there is a good confidential relationship between them.

The counselling model

A well known *counselling model* has been developed by Carl Rogers for psychotherapeutic purposes (16). Rogers considers that people are prone to many psychiatric problems if their self image does not fit the way others react to them. They dare not assimilate these reactions in their own self image, but they can be helped with this problem through use of non-directive discussion techniques. Helpers (in our case, extension agents) show that they understand clients (farmers) and accept what they say about their feelings by repeating them in their own words. We call this 'reflection' or 'mirroring', as it also serves to see if helpers have understood clients. Helpers must not judge what clients have told them, and must not assume they know the reasons for the client's feelings. If clients are vague or unclear about their feelings, helpers can ask for clarification within the clients framework of thought. Where there is a good confidential relationship helpers can confront the clients once with different statements that do not agree with each other.

Non-verbal signals as well as verbal communication from the extension agent are very important for the course of the discussion. Farmers, as clients, will infer the extension agent's attitude towards them from the way the extension agent looks at them, the way he or she sit, his or her tone of voice, etc. It is most important for the farmers to feel the extension agent really is interested in them and in their problems, and is not helping them only because he or she is paid to do so. The feeling that they are accepted as they are by someone who is interested in them can give the farmers strength to accept information which is essential for them to revise their self image.

The counselling discussion technique seldom is enough for an extension agent to help farmers with their decision-making, although it can be a useful way of making the farmers aware of the way their feelings influence their image of the problem. Factual material also is necessary to solve most extension problems. The farmers sometimes have this factual information already, or know where to get it, although they may have difficulty in using the information because of feelings which they are not well aware of. Besides this, the farmers will have more trust in the extension agent and will be more likely to accept his or her information if they feel the extension agent accepts and understands them.

The diagnosis–prescription model

The diagnosis–prescription model often is used by doctors. They ask the patient a series of questions, the sense of which the patient may not understand. Then they diagnose the cause of the problem on the basis of the patient's replies and give a prescription to solve it. Extension agents who have had only a technical training often are inclined to use this model. However, clearly it is unsuitable for problems such as an increase in farm size which have important consequences for the farm family. In the first place, value judgements play an important role in such a decision. On what grounds will the extension agent be justified in making these judgements for the farmer? Furthermore, the choice only will be implemented well if the farmer agrees emotionally with the solution. It is quite probable a farmer will not manage the farm efficiently if he or she decides to enlarge it because of advice from the extension agent.

The participation model

Preference often is given in extension discussions to the participation model in which decisions are made as much as possible by the farmer, although the extension agent may contribute expertise if the farmer appears to need it. By using this model the extension agent can integrate his or her own knowledge with that of the farmer. In doing so the extension agent should bear in mind ideas about decision-making from Section 5.5 when structuring such a discussion. As a rule, discussion will begin along counselling lines in order to get a clear picture of the problem, of the farmers' goals and why they think they have not yet achieved these goals. This may result in redefinition of the problem. The farmers then contribute their factual information in the next phase and, if necessary, the extension agent adds to it. This allows them to see which alternative solutions are possible and what results can be expected from each of these alternatives. It then may appear impossible to reach the goals that have been set, so that the goals must be adapted to the possibilities. Discussion also will be more in the form of counselling when choosing from the alternatives because the farmer is responsible for the choice.

In addition to combining aspects of the counselling model when using the participation model, the extension agent can employ the following techniques:

Asking questions

The farmers can be stimulated by questioning to contribute relevant knowledge and to recognize gaps in their own knowledge and insight, thus becoming more creative. The questioning always is directive to a greater or lesser degree because it will be influenced by the premises and judgements of the questioner. Questioning will work well if the farmers accept the extension agent's premises, if they consider the questions as evidence of the extension agent's interest or if they consider it a challenge to their intelligence. Questions have an advantage over counselling in that discussion moves faster. Rate of progress of discussion will be retarded if confrontation with the extension agent's point of view implicit in the question leads to resistance by the farmers.

Giving information and proposing model solutions

The farmers are confronted with the problem that they must absorb new information from the extension agent into their current thought patterns. They will show interest in this information and in any proposals made by the extension agent if they are introduced when they are ready to absorb them mentally. The information may not be taken up if introduced too early or in an unmanageable form.

Commenting on observations

The extension agent can make a very direct attempt to influence the farmers by commenting on what the farmers have been saying or on their performance as farmers, using their own standards as a reference point. The more their judgements are based on hard fact, the sooner the farmers will be inclined to accept them. An approving judgement strengthens existing opinion even more if the farmers regard the extension agent as an authority. However, the farmers' reactions can be very different with a disapproving judgement. The more uncertain and vulnerable they feel, the more likely that such disapproving comments will cause them to react by escaping or becoming defensive or aggressive, thus reducing the likelihood that extension will help them. Some people with assertive or aggressive personalities need to be confronted strongly with the extension agent's opinion and can benefit from it. However, it is best to avoid such a hard line during first contact if the extension agent does not know the farmer well. Some farmers are convinced that good and evil spirits influence the way their crops and animals grow, a ridiculous idea for some academically trained extension agents. However, if the extension agents express their feelings they are likely to lose the farmers' trust. It will be more effective to create situations in which farmers discover factors that influence their crop growth, such as fertilizers, pest control, land preparation, etc. Their ideas about spirits probably will fade away more rapidly in this way than by direct attack.

Making an inventory of advantages and disadvantages

It is difficult for the extension agent and the farmer to hold informative discussion if they have opposing ideas about possible solutions to a problem. This often leads to a discussion in which each defends his viewpoint but hardly listens to the other. An extension agent can try to prevent this by following the scheme outlined in Table 5.2, making an inventory of the advantages and disadvantages of the alternatives. The agent should begin by questioning the farmer in a counselling manner about what the farmer considers to be the advantages of the alternatives the agent has suggested. The farmer often will be prepared to discuss the disadvantages of the alternatives if he or she thinks the extension agent understands these advantages.

The inventory method can work well if it concerns emotional advantages and disadvantages, but extension agents often work with farmers who are convinced honestly that certain factual information is correct which the extension agent believes is incorrect. In that case it is difficult to get a farmer to change. However, the extension agent must strive to achieve change with important information to ensure wrong decisions are not made. Furthermore, the farmers may lose some faith in the extension agent if they find out later that the extension agent had not drawn attention to their incorrect point of view. The extension agent may do this by creating a situation in which the farmers discover their incorrect viewpoint for themselves. A 77-year-old farmer told one of us that he learned the advantages of using fertilizer at an agricultural course 50 years earlier, but he and his fellow students did not trust these theories at that time. However, his attitude changed when his teacher took him to a plot of oats on his own farm where one part had grown better than another. The farmer said he had not understood why there was such a difference until the teacher explained he had applied fertilizer to the better side. The fact that the farmer talked about his discovery so many years later indicates the strong impression it had made on him.

The farmers may not always discover their own incorrect interpretations of events, and sometimes will have to accept ideas on the basis of the extension agent's authority. This is possible only when the extension agent has built up a strong faith in his professional knowledge and his willingness to serve the farmers' interests, and/or the extension agent has clear evidence that his or her information is correct.

We drew attention when discussing the counselling model to the fact that it may be useful sometimes to confront the farmer with opposing feelings. Such confrontation also can be useful here if the extension agent is convinced the farmer is about to make a disadvantageous decision, and if the farmer is convinced the extension agent has decided to confront him or her because the agent is striving sincerely to serve the farmer's best interests. Let us suppose that a farmer plans to enter a new branch of agriculture such as broiler production, but the extension agent is convinced the farmer will not succeed because of competition from large-scale commercial enterprises. It would help the farmer considerably if the extension agent explains why broiler growing is a wrong choice for him or her and which alternatives are more promising. It would be even better if the extension agent discusses this decision in such a way that the farmer discovers him or herself that broiler growing is not suited to his or her situation. Unfortunately this approach is not always possible because of time constraints.

Where and how to hold mutual discussions

Mutual discussion in agricultural extension can take place in farmers' fields, in their homes, on the telephone, in the extension office or in a teashop in the market place where farmers are accustomed to gather. An important advantage of holding discussions in the farmer's fields is that the extension agent can observe the condition of the crops and animals and diagnose any problems. Also, the extension agent may be able to demonstrate a skill the farmer needs to solve the problem. Discussions in the farm house often take place with the farmer's spouse present. This may save the farmer the trouble of convincing the spouse later of the need for changes. Furthermore, the spouse can become a valuable supporter of the extension agent.

Telephone calls are an important extension medium in industrialized countries because they can save much travelling time. In several countries the Village Extension Agents are available for an hour early in the morning to take telephone calls. A farmer can discuss what needs to be done that day for plant protection or some other very timely activity.

The market place is an important meeting place for farmers and hence is important for extension work. The extension agents can visit the market to talk with farmers they meet there, or they can make regular visits to the same teashop where farmers may seek them out to discuss their problems. There may be little privacy in these public places, but it is important for the extension agent to be readily accessible.

It is difficult to learn how to conduct effective mutual discussion from a book. We must practice these discussions ourselves and evaluate how they have been carried out. We learn more rapidly from our experience with mutual discussions if we receive feedback about the feelings and reactions from our discussion partners. Practical exercises in training programmes provide excellent opportunities to try different discussion methods, especially where participants are encouraged to give each other detailed feedback about their performance.

Some of the theoretical considerations about discussions outlined in this section can be useful in different ways. Firstly, they can suggest interventions to try in discussion, such as reflection. We may see that techniques such as this really do have the expected effect. In the second place, these theories help us to evaluate our own discussion experience, and provide experienced extension discussion leaders with a framework to structure their experiences more thoroughly. Such frameworks or models help people to recognize what they may have been doing for a long time without thinking about it. They then are in a better position to reassess their own performance, to discuss it with others, and to learn and generalize from their experience.

6.3.2 Section summary

Mutual discussion is a very costly but very important extension method. It allows the extension agent to combine a farmer's information with his or her own to solve a unique problem, while at the same time making the farmer aware of the part his or her feelings play in decision-making.

This method requires the extension agent to have considerable skill with personal contacts, and to understand him or herself. The method is especially significant if the extension agent is prepared to help the farmers make their own fundamental decisions.

The extension agent can draw on the counselling model, the diagnosis–prescription model and the participation model when engaging in mutual discussion. The latter model usually, but not always, is preferable. It is reviewed in Table 6.5. The manner in which discussion is carried out, and especially the extent to which the extension agent understands and accepts the farmer, has considerable influence on the confidential relationship between the farmer and the extension agent.

☞ Discussion questions

1 An extension agent can use both group and mutual discussions to help farmers become aware of and assimilate their feelings. Discuss the advantages and disadvantages of both methods. In what situations would you give preference to group discussions or to mutual discussion?

2 An extension agent can look for a solution to a farmer's problem and then advise the farmer to apply the solution. The extension agent also can help the farmer to find his or her own solution. What are the advantages and disadvantages of these two approaches? Use a specific example to show how the extension agent can help a farmer or a group of farmers solve their own problems.

3 A farmer asks an extension agent for information about entering a new type of agricultural enterprise and expects the extension agent to give him or her directive advice. However, the extension agent considers that this problem should be treated non-directively. How can the extension agent overcome the farmer's expectations without upsetting the relationship? (This situation could be role-played, with later analysis of how the extension agent led discussion and why the farmer reacted as he or she did.)

4 Practise a mutual discussion in which one member of the group gives extension advice to another member on a topic in which he or she is 'expert'. Analyse what aspects discussed in this chapter could be identified.
5 Which applications of learning theories did you recognize in our presentation of group and mutual discussions?

Table 6.5 Basic principles of discussion between extension agent and farmer using the participation model

Phase in process	How to conduct discussion
Establishing what the problem is	**1** Let the farmer state his or her own problem. **2** Help the farmer with this by listening carefully. **3** Help the farmer by regularly, clearly and specifically summarizing what he or she has said. **4** Do not underestimate the farmer's problem. **5** Accept feelings as 'objective' facts. **6** Together with the farmer, stipulate what the criteria are for a satisfactory solution. **7** Together with the farmer, search for possible causes of the problem.
Searching for a solution	**8** Jointly look for a feasible solution. **9** Jointly consider all the advantages and disadvantages associated with each solution. **10** Summarize these regularly. **11** Contribute technical knowledge where needed, or draw farmer's attention to other sources (other experts, meetings, publications, etc.). Only contribute this information when the farmer appears to need it but does not have it him or herself.
Choosing a solution	**12** If necessary, jointly ascertain what are the criteria for a good solution. **13** Jointly restrict the number of possible solutions. **14** Discuss in detail the expected results and side effects of these solutions as graphically as possible.
Evaluation	**15** Ask the farmer if he or she is satisfied with the chosen solution. **16** If not satisfied, start again at Stage 6, after allowing time for reflection. **17** Arrange the next step with the farmer at the end of each discussion (new appointment, discussions with others, etc.).

From Oomkes, F.R. (1980) *Handboek voor Gesprekstraining*. Boom, Meppel.

6.4 Media combinations and use of audio-visual aids

'He who does not appreciate school chalk should not be encouraged to use a video recorder.' Heynen

In this section we will discuss how different media can be combined in an extension programme and in what way extension agents can use audio-visual aids to increase common meaning for their message. We discuss ways in which the message can be structured, advantages and disadvantages of some aids and needs and possibilities for research to improve the use of audio-visual aids.

It is clear from what we have already discussed that each extension method has its own advantages and disadvantages. Hence there is no sense in asking which is the best method, although it is sensible to ask what part each method can play in an extension programme. Generally it is recommended to combine different methods purposely. In other words, there is a best method for each purpose or function to be fulfilled. Sometimes different media will be used at the same time; for example, when a lecture is supported with audio-visual aids. Sometimes they will be used in succession, as when written materials are used to prepare farmers for a group discussion.

The UK Open University has used this multimedia approach very systematically. Initially television and radio were to be the main media of instruction. It soon became clear that television was not always the most appropriate teaching medium. Radio is much cheaper and will be as effective when information need not be presented visually. Furthermore, it is often important for students to be able to re-read information. Hence exercise books and written materials are used frequently. Meetings also are held to give students the opportunity to carry out practical laboratory exercises or to discuss issues with a tutor. Personal discussions between teachers and students also are necessary to help students overcome individual difficulties.

This multimedia approach is important not only to perform each vital function in the communication process, but also to perform it through the most suitable medium. In addition, it is usually possible to reach a wider audience. Thus you can reach less educated people with greater ease through television, and more educated people through printed materials.

Advertising research has shown that the person who receives the same message through different media will pay more attention because he or she recognizes something familiar from another context. Although slightly exaggerated, this may be called a shock of recognition.

When planning combinations of media we should think not only of large, well organized campaigns, but also of using audio-visual aids to support talks and group discussions. In this context we refer to aids such as blackboards, flannelboards, flipcharts, overhead and video projectors, photographs, drawings, graphs, maps, slides, filmstrips, film, radio, television, sound and videotapes, CD-ROMs (Compact Disc Read-Only Memory) and programmed learning terminals.

The technical capabilities of audio-visual equipment are being developed very rapidly, although research into their use, effects and role in teaching and learning has not kept up with technical development. Videocassettes, and more recently CD-ROMs, attract considerable attention. For example, CD-ROMs on a disc approximately 12 cm wide and 1 mm thick can carry large quantities of print images as in books and encyclopaedias, as well as colourful illustrations. Some people suggest they will partly replace the book, although there is little objective information to support or refute such a contention. Statements of this nature currently are based on experience rather than scientific research.

6.4.1 Benefits of audio-visual aids

Despite these shortcomings it is clear that audio-visual aids can increase extension effectiveness. For example, an audience is likely to remember more from a talk illustrated with appropriate audio-visual aids than from a speaker who uses no aids. There are several reasons for this:

- aids capture audience attention;
- they can highlight the main points of the talk clearly;
- people remember messages perceived with several senses better than those perceived with a single sense;
- the possibility of misinterpreting concepts is reduced;
- some aids help us to structure our messages in a systematic way.

Software such as Microsoft® Powerpoint® in conjunction with a personal computer permits rapid and easy preparation of materials for projection through video screens or projectors. Whereas one author used to arrive for his lectures equipped with printed notes and many overhead projector transparencies, he now carries only a computer diskette which contains his teaching materials for a semester.

Audio-visual media can serve two different functions:

(1) to improve the process of information transfer (mainly a cognitive process), and
(2) to develop or strengthen the motivation to change (in the first place, an emotional process).

Films and video can be useful in the latter process because they can arouse farmers' emotional involvement with problems the extension agent wishes to discuss. Drama groups and other folk media also can serve this purpose effectively, and can motivate farmers to participate in a fruitful discussion.

6.4.2 Their effective use

The structure of the message must be carefully thought out if audio-visual aids are to be used effectively for information transfer. Hence it is very important to coordinate the audio or sound channel with the visual channel so the reactions to one synchronize with the other. Otherwise the audio signals conflict with the visual, confusing the receiver. Planning must be directed towards uniting word and image. Thus when discussing a film you should first outline its key points, show it, then hold group discussion.

It is very tempting to present a large amount of information in a short time when using different media. This makes it difficult, especially for the less educated members of the audience, to decode and interpret the messages correctly. Hence only limited information should be presented when members of the audience cannot decode information at their own pace.

Audience members will find it very useful to discuss their interpretations of media presentations with an extension agent so the latter can correct any misunderstandings immediately. Although this may not always be possible in practice, it is very important when preparing audio-visual material. Sometimes the audience interprets audio-visual messages in ways the extension agent had not expected, but which are not illogical when you consider their framework of reference. This feedback then gives the audio-visual producer an opportunity to modify the message in order to prevent further misunderstanding (17).

The image of a malaria mosquito projected on a screen from a 35 mm slide may show wings and legs a metre long. Such a picture may not be understood by audiences in remote African or New Guinea villages where inhabitants have not seen a slide projector and are unfamiliar with the idea of photographic enlargement and projection. In much the same way, coloured photos are not always easier to interpret than those in black and white. The number and nuances of the colours confuse viewers and cause the main point to be lost if they are not accustomed to read or interpret pictures. Research has shown that the pictures most easily understood are photographs in which all irrelevant details in the background are blocked out, or three-tone drawings of objects the farmers know in everyday life. No single art style appeared to be best for non-literate people (17). Often there are traditional forms of pictorial representation which could be effective, although they may differ markedly from realistic representations which educated extension workers are most likely to be familiar with.

Audiences nearly always pick up much irrelevant information, so the visual aid producer must try to restrict this by using appropriate forms of presentation. On the other hand the designer must not exclude so much detail that the audience cannot recognize reality. The amount of information included in a picture should determine how long it is shown. A description of the key points and side issues portrayed also may help an audience to understand a complicated drawing or picture. Well educated people or those who specialize in a particular subject often underestimate the time illiterate people need to understand a picture well. It is perhaps even more important to pre-test audio-visual material than written material.

Sometimes the method of presentation attracts so much attention that it detracts from the message itself. Hence it may be necessary to explain first how a flannelboard works and how the cards or figures adhere to it, so the talk is not interrupted by chatter and side discussions about the medium itself.

Films, video and plays can achieve a mixed goal of providing information in a relaxed way. However, we run the risk that relaxation will dominate, and the information content may be missed. Dramatic performances may stimulate an audience to think about a particular problem. However, members must be highly motivated and willing to make a considerable effort to process much information if they are to absorb useful new knowledge.

Use of film and video restrict the extension agent to what is available in a catalogue. Films and videocassettes seldom portray the exact message required by an extension agent, and they cannot be modified to meet his or her needs. However, films can attract large crowds in less industrialized countries. Films and video also are good media for presenting dramatized messages which stir the emotions.

Other audio-visual aids such as a series of 35 mm slides, a flannelboard kit, or even a play or a puppet show can be developed to suit the interests of specific groups, especially when the extension agent produces them him or herself. It also is relatively easy to make small changes to suit different audiences, for example, by changing the photographs in a slide set or the language used on an audio tape where there are many different local dialects and customs. Simple media such as blackboards, flipcharts and flannelboards also present few problems with equipment maintenance, whereas films, videocassettes and computers require relatively costly projection equipment, plus a reliable source of electric power.

Another important advantage of simple media, including manually operated slide shows, is that the extension agent presents and often makes them him or herself and can involve the audience more actively in the subject matter. If the extension agent has some skill at drawing pictures he or she can develop the outline of the topic little by little, highlighting important parts and their relationships rather than presenting the whole picture at once. In contrast with film and television presentations which seldom are interrupted, the extension agent can stimulate questions and provide answers during the presentation with the simple media. This provides the opportunity to relate messages on the audio and the visual channels and to lead the audience step by step to the main points. Local media prepared by the people themselves can be used successfully in participatory approaches.

6.4.3 Evaluating their use

Good evaluation research into the best ways of using audio-visual aids is badly needed, especially into the most effective ways of combining different aids under certain circumstances. For example, currently we have no research findings which can tell us whether and when a flipchart is more effective than a slide series in promoting learning. Furthermore, we need more information about the importance of different sequences of methods or aids, and how each method reinforces another.

Considerable research has been conducted to compare university lectures with television presentations of the same content. Results indicate that neither method is better than the other, although no analysis was made of how the lecturer used television and how he gave a lecture. Clearly, presentation method and channel used should be two different variables in such an investigation. The use of diagrams to explain a relationship is one variable, and the manner in which these diagrams are presented on a blackboard or in a book is another variable.

Extension agents are advised to evaluate their own methods critically. They can use simple pre-tests to check whether farmers understand one type of presentation more easily than another type. Points not clearly understood then can be clarified when redesigning the presentation. A critical approach towards their own methods forces the extension agents to formulate their teaching objectives and specify the composition of their audience very carefully.

Variables to consider

A scientific approach to evaluation research shows the relationships between different variables:

(1) The extension agents' knowledge and their personal skills and interests. (Can they draw well? Do they enjoy public speaking? Are they good discussion leaders?)
(2) The nature of the extension message. (Abstract concepts are difficult to convey visually.)
(3) The type of audio-visual aid. (How much does it cost? How many people can it reach at the one time? An extension film lasting half an hour may cost as much as the annual salaries of three extension agents in an industrialized country.)
(4) The audience composition. (How well educated is the audience? How interested are they in the subject?)
(5) Personnel and finance available. (Knowing that knowledge transfer can be just as effective in some situations with a television programme as with a live lecturer, we could decide to replace oral presentation by a speaker with lectures played back on a video recorder. Folk media are often much cheaper than modern information technology.)

(6) The audio-visual aids available. (Is there a film about a specific subject? Is electric power available? Technically advanced methods often fail in less industrialized countries because of electric power failures, lack of spare parts, poor standards of equipment maintenance, transport problems, etc.)
(7) Ways of acquiring aids. (Is there an efficient distribution system for films or videocassettes?)

All of these elements should be taken into account when programming extension activities.

6.4.4 Section summary

Combinations of mass media and group discussions can bring about substantial changes in behaviour if well organized, especially in less industrialized countries. Many types of audio-visual aids can be used on a smaller scale to increase extension effectiveness. Technical development of these devices is increasing much faster than research into their optimal use. More research attention must be paid to the relationships between variables that influence effectiveness of aids. Messages presented through different aids must be synchronized carefully so the target audience is not overloaded with information. It is essential to pre-test audio-visual materials before distributing them for general use.

☞ Discussion questions

1 Analyse how, in teaching a particular subject you are enrolled in, the lecturer, the textbook, the aids used by the lecturer and student activities relate to each other. Do they complement or support each other optimally? Could other media be used? If so, for what purpose?
2 Choose a topic and a small target group you would like to give extension information to. What audio-visual aids would you use? What do you think you can achieve with each of these aids?
3 What are the advantages and disadvantages of using a video as an extension method for introducing sustainable agricultural practices, compared with the folk medium which is used most often in villages in your country.

6.5 Use of folk media

Interest in use of folk media is increasing in many less industrialized countries. These media include theatre plays, songs, puppet shows, story tellers and other traditional forms of entertainment. It has been suggested that interest in folk media has increased because mass media have been less successful in promoting rural development than was expected 25 years ago, as we discussed in Section 6.1. Alternative means of communicating with rural people therefore are sought. Another reason is that there has been a decreased emphasis on 'top down' communication with rural people in favour of participative approaches. Folk media usually involve substantial participation.

Our current knowledge of the effects of folk media is based more on impressions than on sound evaluative research. These media often appeal to the emotions of members of the audience, who identify themselves with the players. It is easier for them to identify with folk media because these fit closely with local cultural patterns, whereas the mass media bring information from urban centres and from foreign countries. Hence, folk media often are more effective at arousing the motivation to change than in transferring knowledge to villagers about how to change. Furthermore, the audience tends to have confidence in the message because communication takes place in a form they are accustomed to, in their dialect and from people they often know. It can be acceptable for players to ridicule people who prefer to stick to the old ways, whereas it would be unacceptable for a government extension agent to do the same.

Village people may participate in these folk media in different ways. Some can write scripts for plays, some can compose songs, while others can act or assist with the production. The players usually have considerable freedom in the way they perform their roles, both in terms of the words they use and their non-verbal communication. Audiences likewise are seldom passive listeners or viewers, but tend to participate actively, thus influencing the performers. Furthermore, those who convince villagers, say, of the need for soil conservation, usually become convinced themselves. Extension organizations can stimulate participation by arranging competitions for the best play or song which promotes adoption of desirable innovations such as improved rice production or the formation of co-operative societies. Extension services also can assist groups which like to perform plays by providing material for their costumes or by arranging transport to neighbouring villages. Although this may appear to be a relatively inexpensive extension method, budget regulations often do not permit its use. The village elite also may play a major role in participative processes associated with development and use of folk media. Unfortunately, they may try to serve the interests of their own group where there are conflicting interests in a village.

Village people do not always accept innovative messages sent through traditional media. The techniques may fail badly unless used by people who thoroughly understand local culture. Cases have been reported where audiences walked out angrily, as a Roman Catholic might do if psalms were changed to promote abortion. There are many other cases where these media have been used successfully. For example, *wayang kulit* or shadow puppets have been used to promote agricultural innovations in Java, while plays written by a village headman have been used to promote the acceptance of soil conservation in the Indian Himalayas.

Traditional media may compete for attention with modern media such as radio, television, cassettes or video, and in some instances have lost the battle. People do not like to be considered traditional in their use of media, but in some situations different media forms can co-exist with each other. The extension service in the Netherlands could expect five times more people to attend a play presented at a farmers' union meeting than would attend a talk about the same topic. Many of the less innovative farmers, and especially many women, will come to a play but would not come to hear a talk. Also, villagers in less industrialized countries often come in large numbers to and are more involved with the performance of a folk medium than they are with modern urban media.

Folk and modern media do not have to compete, but may complement each other. Thus, folk media can be shown on television, and folk songs with a development theme can be broadcast on radio. This often is done in Indonesia, partly because the government sees it as a way of promoting national unity. A Brazilian agricultural extension service commissioned a popular folk singer to write and perform folk songs which incorporate technical messages about fertilizer use.

6.6 Use of modern information technology

We have stressed that information is an important resource in modern agriculture. The development of computers and improvements in telecommunications offer farmers many new opportunities to obtain technical and economic information quickly and use it effectively for their decision-making. Modern farmers are business managers who try to grow the right crops and animals in the most profitable way. Farming for them is no longer just a way of life, but a business in which they can earn a good income if they use all the opportunities their environment offers, but in which they may not survive unless they manage their farms more efficiently than most of their colleagues.

The amount of information farmers can and should use for their management decisions is increasing rapidly. It includes reports of research findings, markets, and data on growth and management processes on their own and other comparable farms. They can use this information to select the most profitable production technology, to create optimal growth conditions for their plants and animals, to budget expenditure and see which enterprises are the most profitable, and to decide when and where to sell their products.

Previously they kept accounts for fiscal reasons such as income taxation. Results often were reported to them up to six months after the end of the financial year, too late for them to use for correcting their mistakes. Now rapid or even immediate feedback is possible.

Modern information technology can give the farmers rapid access to a large amount of information, help them select from this exactly the information they need for their decision-making, and with the assistance of decision-making models, guide their decision-making. Modern computers have become much more 'user friendly'. That is to say, the user does not need to know how to program them but only has to answer questions displayed on the screen by pressing buttons.

Previously the mass media gave generalized advice to farmers, but with modern information technology extension can provide advice custom tailored for each farm and farmer without visiting the farm personally. This makes it possible to combine some of the advantages of mass media with those of farm visits. For the past 30 years the price of computers has decreased at about 20 per cent a year, while at the same time telecommunication charges also are decreasing rapidly, although in many countries the costs of extension agents are increasing. Therefore extension managers have to consider seriously in which situations and for which purposes it is desirable to replace personal contacts between farmers and extension agents with information technology. Costs are not the only consideration when doing this. There are also situations in which the communication technology can do a better job, because it provides more or more up-to-date information, provides this information faster and makes it possible to combine internal information from observation on the farm with external information.

However, many attempts to introduce information technology in agriculture have been less used and had less effects on improving farm management than the proponents of this technology had expected. One reason is that information which experts believe is good for farmers has often been provided rather than help with searching for and processing information for which farmers feel a need. The methods farmers and extension agents have to follow to use this technology also may be more complicated than is necessary, because they are not adjusted to their normal way of information processing. Experts often believe that a lack of certain kinds of information creates a major problem for farmers, whereas for many it is a major challenge to use all the information they get from many different sources in their decision-making process. The situation is made even more difficult by the contradictions which often exist between different pieces of information.

Information and communication technologies are developing rapidly. Many new technologies at present not on the market will be available in a few years from now. We can mention here only some of the major kinds of technologies which are used already in agricultural extension.

6.6.1 Electronic data base access and search systems

These data bases may contain information such as characteristics of plant varieties, plant and animal diseases and possible control methods, data and formulas to be used in the design of an irrigation system or to calculate a feed ration, market prices of inputs and products in various markets, weather forecasts, library catalogues and documentation systems. It is time consuming and expensive to keep such data bases up-to-date.

It is possible for a farmer to gain access to many such databases using an equipment combination of a home computer fitted with a modem linked by domestic telephone lines to a remote computer or a CD-ROM. There are also combinations using normal home television sets fitted with a key pad which may be cheaper. The computer gives access to a much wider data base which can be used interactively. For example, it is possible to use data on feeds and animals available on the farm, together with commercial data on feeds and prices and a formula from this database to calculate optimal feed rations for that farm. This is also possible with a CD-ROM attached to the home computer, although such a CD-ROM has to be replaced regularly by an updated version, which may be impractical due to budget constraints. The CD-ROM method may be the best solution in countries without a good telephone system.

Another and much simpler way used successfully in many countries is to install an automated telephone answering system which gives up-to-date information on weather forecasts, prices or plant protection measures required for a certain crop. This has the advantage over the computerized system in that there is a much lower investment in capital and in training. The farmer has only to know which telephone number to call.

6.6.2 Feedback systems

For a long time farmers have used accounts, milk records, etc. as a feedback system. Modern technology makes possible much faster and more effective feedback. For example, sensors can measure how much milk per cow is produced today. Computer models can predict how much could be expected when taking into account the amount which was produced yesterday, the age and the period since calving and gestation for all cows, and the animals which are no longer milked. If this prediction deviates from the actual quantity, it is a warning sign that something has gone wrong. Like a fever thermometer, it does not say what is wrong, but only that one should look for the causes. The earlier they are discovered, the less harm they will do.

6.6.3 Advisory systems

Advisory systems can include decision support systems and expert systems. Simulation models are rapidly gaining importance in research on plant and animal production. Linear programming and similar techniques also are now important in economic research. Application of similar systems at the farm level can give a considerable increase in the level and quality of production and in profits. One Dutch consultancy each week receives data on 100 parameters of rose production from their clients' farms, analyses these data in their computer simulation model, and advises the growers on use of fertilizers, plant protection measures, climate control, etc. It pays to employ one staff member in the consulting firm for each 10 growers. Similar but simpler systems are widely used in plant protection and irrigation management. Information in these systems serves as a basis for discussion between the farmer and the extension agent. However, other systems are more often used in this way than is intended by their designers.

6.6.4 Networks

Farmers in many countries, especially under rain-fed conditions, do not live close to each other but still like to know what problems other farmers face and how they are solving them in other districts. Modern information technology to some extent allows widely separated farmers to maintain contact, irrespective of distance. For example, some Australian farmer groups which participate in the national Landcare programme (13) have joined LandcareNET, a commercial electronic network (18). Farmer groups as well as staff at research institutes and other government agencies can place information on the network, ask questions of other groups, send electronic mail to each other and gain access to international data bases through the Internet. For example, a farmer group in North Queensland was able to locate supplies of a salt tolerant grass species in Western Australia, several thousand kilometres away, by placing a request on the appropriate 'Bulletin Board' or 'conference'.

Table 6.6 gives an overview of the characteristics, advantages and disadvantages of these information technologies. It is quite important when planning to use these information technologies to analyse first how the farmers obtain their information at present, how they use this information in their decision-making and what difficulties they experience in these processes. It is possible to design a prototype technology which can be improved on the basis of its users' experiences. This implies that the technology should not be developed solely by computer experts, but in cooperation between experts, extension agents and farmers.

When farmers use this information technology the role of extension agents changes from one of teaching farmers which decisions to make, to teaching them how to make decisions. For example, in pest control the extension agent used to tell farmers which pesticide they should use. Now this information is available readily from a CD-ROM or computer network, the agent has the opportunity to teach farmers the basic principles of epidemiology on

Table 6.6 Functional characteristics which are associated with various types of communication technology

	Search and access systems	Feedback systems	Advisory systems	Network (transaction) systems
Common labels used in practice	❑ databases ❑ teletext ❑ videotex ❑ hypertext	management information systems	❑ decision support systems ❑ expert systems ❑ knowledge systems	❑ E-mail ❑ electronic conferencing systems ❑ videotex
Underlying intervention goal	provide efficient access to information	provide adequate feedback	give specific support and advice	facilitate networking activities
Means	search/selection procedures	registration, manipulation and representation of data	calculation, optimization, simulation and reasoning	message exchange, file-transfer (pictures, sound, text, etc.)
Source of information	information suppliers	end-users	end-users and 'experts'	end-users and information suppliers
Aspects of learning, decision-making and problem solving	mainly image-formation and implementation	mainly evaluation, image-formation and problem identification	mainly searching/selecting alternatives and image-formation	variable: dependent on nature and content of messages
Role of communicative intervener	supplier of information	discussion partner for the interpretation and processing of information	❑ discussion partner ❑ co-user ❑ corrector	user
Typical substantive problems	search procedures are incompatible with knowledge and logic of end-users	feedback does not (or no longer) meets the interests of the end-users	❑ validity of the underlying model is questionable ❑ complexity causes interpretation problems	users are confronted with superabundance of information

From Leeuwis, C. (1996) Communication technologies for information-based services: Experiences and implications. In: The Contours of Multimedia. Recent technological, theoretical and empirical developments. Academia Research Monographs 19, John Libbey Media, Luton, pp. 86–102.

which the computer model is based. They might learn, for example, that a crop infection at the end of the growing season will have little impact on yields so that it does not pay to spray. The problem remains that the farmers may not dare to apply the computerized recommendations if they do not understand the underlying principles, and they do not know whether they have to deviate from these recommendations because the assumptions on which the model is based are not valid for their situation.

The extension agents also can help farmers by teaching them:

- how to select a computer and a computer program;
- which data they have to collect and record on their farms to use with the computer program;
- how to collect these data, for example, how to recognize the infection rate of different wheat diseases;
- to select the information they need for their decision-making; and
- how to interpret correctly the information they receive.

Extension staff will require special training in how to prepare material for these new electronic systems.

Farmers' need for modern information technology depends on the sophistication of their decision-making and on their willingness to collect and record the data required for this decision-making. Unsophisticated farmers can obtain all the information they require in the traditional way. To date it appears that this new technology is used less frequently than was expected by the people who developed the information systems. This is partly because farmers first have to learn how to use the technology correctly and partly because many older farmers have resisted learning the keyboard skills. It takes time to convince them that computer models generate reliable data for their situation. Some of the early models made rather unrealistic assumptions and were not really geared to farmers' information needs. They were based on a sound knowledge of growth processes rather than on an analysis of the information farmers need for their decision-making processes.

The move towards *expert systems* is an important development in information technology. These advise the farmer which alternative to choose from a wide range of possible alternatives by processing data from a large number of variables according to certain decision rules. These systems can apply the decision rules more consistently and process the relevant data more effectively than the farmer can himself.

Expert systems and other simulation technology require a high level of agricultural research with the capacity to develop relevant data and models to process these data. The technology can have significant impact on income in an enterprise such as glasshouse horticulture where large savings can be made in energy required for heating. Production levels can be increased by making intensive use of information on environmental conditions which influence growth processes. There are fewer possibilities for increasing income from field crops in this way.

Information technology makes farmers more dependent on companies and agencies which supply and maintain the hardware and software. This may be not only expensive but also dangerous if the support is unreliable or given without the best interests of the farmers in mind.

The temptation to use new technology because it is available has led to some failures. It is sounder practice to analyse the communication problem first, then look for the most suitable technique to solve it. This could involve use of the most advanced technology, but it may very well be an ancient technology such as the use of itinerant singers.

6.6.5 Section summary

The technical possibilities of information technologies are increasing rapidly, but so far they have not been used very successfully in extension. One reason for this slow adoption is that the design of these technologies has been more sender than receiver oriented. Little attention has been paid to analysing difficulties experienced by farmers in use of the technologies, and investigation of how these difficulties can be overcome. Unfortunately, information technologies seldom are developed in association with the people who are expected to use them.

6.7 Chapter summary

Instead of a summary we offer Table 6.7 which reviews some of the advantages and disadvantages of the extension methods discussed. It will be clear from our earlier discussion that these advantages and disadvantages depend

Table 6.7 Functions, advantages and disadvantages of different extension methods

Medium is suitable for or has the characteristic	Mass media	Talks	Demonstrations	Folk media	Group discussions	Dialogue
Creating awareness of innovations	xxx	x	xx	xx	0	0
Creating awareness of own problems	0	x	xx	xxx	xxx	xxx
Knowledge transfer	xxx	xx	xx	xx	x	xx
Behavioural change	0	0	xx	x	xxx	xx
Using other farmers' knowledge	0	0	x	xx	xxx	x
Activating learning processes	0	0	x	xxx	xxx	xx
Adjustment to farmers' problems	0	0	x	xx	xx	xxx
Level of abstraction*	xxx	xx	0	0	x	x
Cost per farmer reached*	0	x	x	xx	xx	xxx

0 = unsuitable. The number of crosses indicates suitability, except where marked with an asterisk which indicates the level of abstraction or cost.

not only on the method chosen but also on the way the method is used. The nature of the audience and of the message receive only limited treatment in the table. Many of the points outlined in the table are based on impressions rather than on established research results.

We have chosen a different approach in Table 6.8 by indicating which educational goals can be achieved with which methods.

Table 6.8 Strategies and methods to reach different learning goals

Nature of learning goal	Strategy	Preferred methods
Knowing (cognitive)	Transfer of information (from outside)	Publications and recommendations in mass media, lectures, leaflets, directive dialogue
Attitudes (affective)	Learning by experience (information from inside)	Group discussions, non-directive dialogue, simulation, certain types of films
Action/doing (psychomotoric)	Exercises in skills	Methods which encourage action = training, preparation by demonstration or demonstration films

Guide to further reading

General

Blackburn, D.J. (ed.) (1994) *Extension Handbook: Processes and Practices*. Thompson Educational Publishing, Toronto.

Mass media

Practical suggestions for the use of various communication methods are given in:

Anonymous (1987) *Communicating Development*. International Institute of Rural Reconstruction, Silang.

Cherry, G. and Harvey, N. (1980) *Effective Writing in Advisory Work*. Ministry of Agriculture, Fisheries and Food, London.

McQuail, D. (1994) *Mass Communication Theory: An Introduction*. Sage, London.

Nair, K.S. and White, S.A. (eds) (1994) *Perspectives in Development Communication*. Sage, New Delhi.

Windahl, S., Signitzer, B. and Olson, J.T. (1991) *Using Communication Theory: An Introduction to Planned Communication*. Sage, London.

Woods, R.A., Hixson, P.C. and Evans, J.F. (1994) Using Communication Media. In: *Extension Handbook: Processes and Practices*. (ed. D.J. Blackburn). Thompson Educational Publishing, Toronto.

Public speaking

De Vito, J.A. (1994) *The Public Speaking Guide*. Harper Collins, New York. An expanded version of Units 19, 20 and 21 in De Vito (1994) *Human Communication*, 6th edn. Harper Collins, New York.

Demonstrations

Krishan, R. (1965) *Agricultural Demonstration and Extension Communication*. Asia Publishing House, Bombay.

Discussion groups

Bertcher, H.J. (1994) *Group Participation; Techniques for Leaders and Members*, 2nd edn. Sage, Thousand Oaks.

Chamala, S. and Mortiss, P.D. (1990) *Working Together for Landcare: Group Management Skills and Strategies*. Australian Academic Press, Department of Primary Industries, Brisbane.

Hartley, P. (1993) *Interpersonal Communication*. Routledge, London.

Individual extension

Brammer, L.M. (1973) *The Helping Relationship*. Prentice Hall, Englewood Cliffs.

Egan, G. (1993) *The Skilled Helper: A Model for Systematic Helping and Interpersonal Relating*, 5th edn. Brooks Cole, Pacific Grove.

Heron, J. (1990) *Helping the Client: A Creative Practical Guide*, 3rd edn. Sage, London.

Audio-visual aids

Hoffmann, V. (1991) *Bildgestützte Kommunikation in Schwarz-Afrika: Grundlagen. Beispiele und Empfehlungen zu angepaszte Kommunikationsverfahren in ländlichen Entwicklungsprogrammen*. Margraf, Weikersheim.

Minor, E. & Frye, H.R. (1977) *Techniques for Producing Visual Instructional Media*, 2nd edn. McGraw-Hill, New York.

Werner, D. and Bower, B. (1982) *Helping Health Workers Learn: A Book of Methods, Aids and Ideas for Instructors at the Village Level*. Hesperian Foundation, Palo Alto.

Folk media

Valbuena, V.T. (1986) *Philippine Folk Media in Development Communication*. Asian Mass Communication Research and Information Centre, Singapore.

Information technology

A good analysis of the potential of communication technology is given by:

Woods, B. (1993) *Communication Technology and the Development of People*. Routledge, London.

An excellent discussion of the potentials and difficulties of using information technologies in extension is given by:

Leeuwis, C. (1996) Communication technologies for information provision and advisory communication: Experiences and implications for multimedia development. In: *Multimedia: A Critical Review of the Technology and its Applications* (N. Jankowski and L. Hanssen). Libbey Media Publications, London.

Studies on the use of Information and Communication Technologies are discussed in:

Beers, G., Huirne, R.B.M. and Pruis, H.C. (eds) (1996) *Farmers in Small-scale and Large-scale Farming in a New Perspective: Objectives, Decision Making and Information Requirements*. Onderzoeksverslag (Research report) 143, Agricultural Economics Research Institute, The Hague.

7 Planning Extension Programmes

'If you have no goal, then any road is the right road.' Koran

7.1 Introduction

We have shown in the previous chapters that many decisions have to be taken in order to ensure effective extension work. Naturally these decisions must be in tune with each other, which implies that extension requires systematic planning. Hence, planning an extension programme involves decision-making about the work of an extension organization. A programme is required for short activities, such as giving information about new rice varieties to a farmers' meeting, as well as for activities which might occupy a group of extension agents for several years, such as doubling rice yields per hectare through use of improved production technology.

Our analysis of the programme planning process is based on Bos' decision-making model, which makes a distinction between the choices which have to be made and the knowledge which has to be gathered to make these choices. (See Section 5.3.) Extension agents have to decide, either by themselves or jointly with others, about:

- the goals they are aiming at;
- the target group they wish to help with opinion formation and/or decision-making;
- the content of the extension message;
- the extension method or combination of methods to be used, and how they will be used; and
- the organization of all activities.

On the pathway towards knowledge extension agents need information about:

- the goals;
- their target group;
- the behavioural alternatives for the target group, and the results expected from these alternatives;
- the media they use or could use; and
- additional resources available to them.

There are two major difficulties when planning an extension programme like this. In the first place, the five decisions mentioned above will influence each other. For example, if different goals are selected then usually each one of the other decisions will change also. In the second place, an extension programme tries to change farmers' behaviour either faster or in a more desirable direction. However, whether or not these objectives are achieved depends mainly on the farmers rather than the extension agents themselves.

In this chapter we will attempt to show how to overcome both these difficulties. We emphasize that it is an attempt, as there is little empirical research evidence to support our suggestions. Hence, at the end of the chapter we have indicated which research needs we consider urgent. We begin the chapter by discussing pathways towards gaining knowledge; then we consider how we choose from one of the five decisions given earlier. Finally, we propose in which way these decisions can be taken, paying special attention to the role farmers play.

Before we commence our discussion we should point out that many extension services do not work with systematically planned extension programmes. Much of their time is spent reacting to farmers' questions. They may also work incidentally with mass media, sometimes on their own initiative and sometimes because of requests from an editor. This approach has the advantage that extension agents can respond easily to requests for information from those needing help. It also means that the service invests little time in planning an extension programme. This type of *reactive extension work* is necessary, and time should be reserved for it in the extension programme. We know from past experience which questions will be asked most often and we should be ready to answer them correctly with minimal effort. This can be achieved by training staff how to answer these questions, by preparing extension materials and by mass media messages and farmer training programmes which deal with a number of these issues. For example, this approach might enable farmers to recognize the major plant diseases themselves and to find in a booklet how they can be treated.

There is a danger in reactive extension work that new problems will be identified too late. 'Prevention is better than cure' may be ideal, but preventive work is comparatively rare outside livestock disease prevention. Extension advice usually is sought when the problem already is present.

Despite the importance of reactive extension work the main aim of an extension programme is to *initiate change*. The programme indicates how extension can contribute to agricultural development. The extension service has to start activities to initiate these changes because many farmers do not ask to help when it is needed most. They do not have sufficient insight into their own problems to formulate a clear request for help. Furthermore, extension agents tend to give most help to those farmers with similar views to their own. Farmers often ask for help with problems for which they have been educated themselves and hence know what questions to ask. Also, they are more inclined to ask for help with minor technical problems than with major management problems. They believe a good farmer should be able to solve these management problems himself. Farmers sometimes fail to seek advice because they are unable to diagnose the cause of a problem and therefore underestimate its importance.

Extension agents who do not programme their work have to divide their attention between many different problems, so they are unable to pursue any one problem in depth or to participate in additional training to be able to solve each problem effectively. Their motivation can be improved if they know what is expected of them.

7.2 The pathway towards knowledge

In Section 5.5 we pointed out that good decisions require us to travel frequently from the pathway towards knowledge to the pathway towards choice and back. In order to avoid confusion we now will discuss these pathways separately in relation to decision-making about the extension programme.

On this pathway extension agents can either collect the information needed for planning the extension programme in close cooperation with farmers, or they can try to gain the knowledge by themselves or with the help of experts such as a market research bureau. The preferred method depends on the situation and the kinds of problems they tackle (see Chapter 9).

7.2.1 Goal

In Section 2.2 we pointed out that an extension programme usually is one of the policy instruments within a larger development programme. Therefore we need to understand the whole programme, and especially the way in which other different policy instruments are combined with the extension programme. What subgoals does the development programme have which can be activated through education and communication? As a rule these subgoals will become the extension programme goals. For example, the development programme also may use agricultural research, irrigation, supply of inputs and credit, pest control and marketing procedures to increase rice production. We would have to ask exactly what the role of extension is with regard to these policy instruments. Some expectations regarding this role may be unrealistic, but this must be made clear in order to be able to plan a successful development programme.

A methodology to analyse how the extension programme relates to the overall development goals has been developed for German Technical Assistance Projects, but is also used widely elsewhere. The development goals and purpose of the project, that is, its intended impacts and anticipated benefits, are decided in this process. From these are derived the results and outputs of the project, and the activities through which it can be expected these results will be achieved. Objectively verifiable indicators are sought at the level of the overall goal, the project purpose, the results and outputs, and the activities (1).

7.2.2 Target group

Target group analysis is one of the important methods for deciding which problems an extension programme will be aimed at. We also speak of 'needs assessment' which relates on the one hand to the needs felt by the target group and on the other to the needs for change recognized by the extension agents or their organization. Information for this must come from integrating points 1, 2, 3 and 7 below. Target group analysis also is important for selecting the most effective communication methods and content. This analysis should provide information, especially on the following points:

(1) What is the current behaviour of the target group in the area in which the extension organization operates? Special attention should be paid to those aspects in which the organization wishes to promote changes because farmers have asked for them, or because extension agents believe farmers will be in a better position to achieve their goals because of the changes. We are concerned here not only with the behaviour itself but also the reason why the target group behaves in this way and the results they expect from this behaviour. An extension organization wishing, for example, to promote marketing cooperatives will have to know how farmers currently market their produce, what problems they have with marketing, their experiences with cooperatives and how they feel about them.

Farming systems research can give valuable information about farmers' current behaviour and the reason for it. This research focuses on interrelationships:

- between different aspects of the farms, such as crops and animals;
- between farms in the community, such as exchange of labour between large and small farms; and
- between the farms and their environment, such as marketing and credit.

These interrelationships clearly are influenced by the resources different groups of farmers have, such as land and capital, and by their access to services from government, cooperatives, private business etc.

(2) What do target group members consider to be their problems? The group will be especially interested in extension about these problems. It is equally important for extension agents to know the areas in which group members feel they have no problems. It will be more difficult to attract farmers' attention if extension agents wanted to offer extension programmes on such issues. For example, a soil conservation officer who is convinced that burning shrubs in the forest is a major cause of soil erosion, should have a different extension strategy when the rural people already are looking for an alternative way of clearing than when they are not.

(3) What knowledge, skills and attitudes do the target group members have towards what they consider to be their problems and towards the problems for which extension agents consider extension help is desirable? Extension agents must find out if certain knowledge and skills for problem solving or carrying out specific activities are lacking, incorrect or deficient. They also must understand what forces are at work towards the proposed direction of change, as well as those opposed to it.

- Are these forces influenced by the target group's experience with past changes?
- Do group norms also play a role here?
- How much freedom does the target group have to diverge from these norms?
- What is the relationship between the attitude of the target group towards the problems and the goals that it strives for?
- Is it possible that the target group strives for different goals from the extension agent, or perhaps for goals other than those the extension agent assumed they strived towards?

(4) Which members of the target group make which decisions, and who influences these decisions? Extension agents in less industrialized countries do not always take into account the division of labour between men and women in these countries. For example, they might underestimate women's influence on agricultural decisions. A good extension programme requires clear insight into target group leadership and willingness of leaders to cooperate with the extension agents. It is also important to know the consequences of a decision for different members of a target group.

(5) Which communication channels do target group members use now and which ones could they use in the future? What is known about this from the evaluations of previous extension programmes? What language or dialect is used by the target group? It is very important to possess this type of information if extension agents are to reach the target group. For example, they should not only know how often target group members watch television but also the types of programmes they watch. The extension agent must decide which channel to use as well as how to use that channel.

(6) How do members of the target group relate to the extension agent? Do they trust his expertise and objectivity? Their attitudes towards him will be influenced heavily by their past experiences with him and his colleagues. Who else is trusted by the target group and could act as intermediary between the group and the extension agent? Such a person may be an opinion leader or hold a leadership position in local society, for example, as a teacher or member of boards or associations. With which other actors in the AKIS do farmers interact on this topic?

(7) What resources do different categories of the target group have access to? Are there possibilities for obtaining more resources? If there are, which ones?

(8) What is the target group's situation and how does this influence target group behaviour? We have pointed out already that extension agents must try to not only help farmers to change but also assist them to improve their situations. Hence the extension agents must be well aware of the farmers' situation and try to predict what the results of these changes will be. This requires a clear view of power relationships such as the role of moneylenders. The problem with such an analysis is that relationships of this kind change frequently. The situation will be very different in a few years time without any intervention by the extension agent. The agent aims to speed up the change process or to change the direction somewhat. Therefore he or she must know not only the present situation but also in which direction it is changing and why it is changing in this way.

Present reactions of the target group can be explained in part by past experiences. For example, resistance to the introduction of birth control measures in many less industrialized countries is caused partly by the high rates of infant mortality in the past. Parents liked to have a large number of children to ensure they would be cared for in their old age. Now that rates of infant mortality have declined, the values and norms that influenced the high birth rate have not adjusted immediately to the changed situation. Such insight into the recent history of a target group can help the extension agent to decide which values must be reconsidered by the target group in order to be able to use their resources optimally in the near future.

(9) What differences are there in the target group towards variables that can be used to segment the group into different but more homogeneous sub-groups which can be given the same message through the same communication channels? In an extension programme concerned with new varieties, for example, it can be necessary to approach farmers who do not yet know about the new variety, but who would use it as soon as they are informed about its advantages, in a different way to farmers who know about the variety, but who do not yet use it because they lack the cash for pay for the seeds. Important variables of this type include the need felt for extension about certain problems, a similar problem situation, knowledge, attitude and

behaviour towards certain innovations, use of communication channels, trust in the extension agent and level of education. Results of adoption research can be useful here. People who have adopted innovations at a certain rate in the past often will be inclined to do the same in the future. It may be necessary to develop different recommendations for different segments of the target group, especially those segments which differ in the resources available, such as farmers with small or large enterprises, male and female farmers, and rain-fed and irrigation farmers. Extension agents also can seek people who soon will have to make similar decisions about similar problems and who are more or less in the same situation. Farmers' phase in the life cycle also can be important for this segmentation.

(10) What are the descriptive variables, such as education, age, gender, socio-economic status, sources and importance of non-farm income and locality?

How can an extension agent obtain this information? There are different methods which complement each other. We can use existing studies of target groups, discussions with people who know the target groups, or make special studies of the target group by talking to group members or their representatives.

The Government Statistical Office has many statistical facts available on a geographical basis about farms, and sometimes also about farm families. Many of these statistics are useful for planning extension work. There also has been much market research in many countries, because industrial planners often consider it irresponsible to promote the sale of a certain product without having adequate information about potential buyers. Naturally the results of most of these investigations are kept secret, although sometimes part of the information will be made available to extension agents. Many governments commission market research projects before introducing new policies, and much of this information is available if you know where to look.

Many countries have published readership surveys of various newspapers and journals to give potential advertisers an idea of the value of different media for their advertising campaigns. Such information also helps extension agents to choose the best channels.

More qualitative but often no less valuable information about the target group is available from many sociological and anthropological studies and from regional novels.

Extension agents commencing work with a new target group often can find people who have regular contact with the group and enjoy its trust. Such people are important sources of information, although at times you have to consider their information more as an hypothesis than as proven facts, as is also the case with many regional novels. It is important first to establish how much contact these people have. Sometimes local doctors, priests, influential farmers, hotel owners and others contact only a select part of the target group without realizing it.

It may be possible to obtain only part of the information required for planning a good extension programme from these sources. An extension organization could let a random selection of target group members fill in a questionnaire or interview them to obtain additional information. Textbooks about social research methods may give some useful leads to obtaining information through research institutes.

The extension agents also should do their own simple research, because they can learn much more from direct contact with the target group than from reading a research report (2). They must pay attention to random sampling so that they obtain a clear picture of the whole target group, rather than of those members who talk regularly to extension agents. The remainder of the group often has other problems and may see them differently. It may be difficult for some extension agents to change their role from someone who teaches to someone who listens, especially with their own farmers who also expect them to give information. It may be difficult for farmers to give reliable answers to questions from extension agents or research workers, either because they know little or have a hazy opinion about the topic or because they think they ought to answer like 'good' farmers or tell the interviewer what they think the interviewer wants to hear. This problem may be reduced by avoiding general questions and asking such specific points as what happened in a particular field in one cropping season. An interviewer may obtain the best information by not following the exact wording of a questionnaire but by formulating questions based on the previous answers given by the farmers or members of their family. Exact recording of answers can give information about the farmers' attitudes. This approach requires capable interviewers who understand which information could be valuable for planning a good extension programme.

Farmers also are more likely to give reliable answers if they can recognize the potential benefits in helping the extension agent and do not regard the collection of information as an examination. The extension agents can supplement and check the information they obtain in interviews through systematic observations of the behaviour of farmers, their crops and animals.

Information collection should be arranged collaboratively with farmers. This helps to create a more equal relationship between extension agents and farmers. Farmers then find it easier to discuss their own problems and to give suggestions for improving extension work. Furthermore, it helps farmers to understand their own situation and that of their group, thus motivating them to work for improvements. It is not unusual for farmers to perceive their problems in a different way from extension agents. Therefore the farmers may not always pay attention to the information agents give them for solving their problems. Effective communication is possible only if farmers and agents perceive the problem in the same way. This is achieved best by farmers and agents analysing the problem together, and preferably also jointly testing solutions to the problem. Such a process of data gathering is at the same time a strategy of change, even though it may be difficult to involve many farmers.

7.2.3 Alternative behaviour

We also require information about alternative behaviours and their expected consequences from which the target group can select possible solutions to their problems. We discuss the point in Section 7.3.4 and in Chapter 10 when considering the management of an extension organization. Hence we will not discuss it further here.

7.2.4 Media

Extension agents need some insight into the expected effects of a medium or channel if they are to make effective choices. As is mentioned in Chapter 6 and in many more detailed publications, they will have to use general information about media effects that is available. Furthermore, they will have to make use of specific information about the target group, as is mentioned in point 5 of Section 7.2.2.

7.2.5 Resources

The resources available will influence the type of extension that can be given. Factors include the number, quality and interests of the extension agents, organization and persons with whom they can cooperate and the means available for transport, for example. There may be opportunities to increase available resources by giving refresher courses to extension agents or by acquiring extra funds for special campaigns.

In the long-term the way in which extension is given will influence resources available. Farmers can influence financial sources to make more funds available, or they may be prepared to meet expenses themselves. The opinions of those responsible for the available budget naturally are very important. Their opinions may be influenced by the way in which extension is given, as well as by public relations activities about this extension work. There are limits to the possibilities extension organizations have to influence the available resources, but extension administrators could make better use of systematic public relations.

7.2.6 Is the situation known sufficiently?

Analysis of the situation described in the paragraph above should make it possible to choose clearly where the emphasis of an extension programme should be placed. We can judge whether or not we have been successful in this by asking the following questions:

(1) Is the present situation described clearly and definitely? Does it give the most important facts? Do these facts indeed demonstrate that it concerns an extension problem?
(2) Does it appear from this description of the situation that there is a definite problem, so that there is a difference between the exact, and if possible, quantitative description of the existing situation and the equally exact description of the desired situation?
(3) Have farmers and their organizations expressed interest in solving this problem?
(4) Is it clear why the desired situation has not yet been achieved? Is this a problem which can be solved through agricultural extension or does it require other solutions such as improved supply of inputs as well?
(5) Has the problem been discussed with specialists in that field and with representatives from the target group?
(6) Can the problem be solved with available manpower and resources?

7.3 The pathway towards choice

7.3.1 Selecting programme goals

We must bear in mind the following when choosing extension programme goals:

(1) Our perception of the present situation.
(2) Our vision of the desired situation.
(3) Our perception of why the present deviates from the desired situation.
(4) What possibilities we see for bringing about changes through extension with available resources and manpower.

We will now discuss these four factors.

We pointed out in an earlier paragraph that extension agents can get an idea of the present situation from farmers.

The question about what we wish to achieve with extension finally rests on value judgements. We discuss the important point of whose value judgements will influence this choice further in Chapter 9.

Ultimately it concerns questions such as 'Is it desirable that less industrialized countries adopt technologies and other elements of industrialized countries? or 'How important do we consider a high income in the short-term versus maintaining the potential for agricultural production in the long-term?'. The problem here is that we often strive for different goals at the same time which cannot be achieved together. For example, there are many people in Holland and Australia who wish to avoid the disadvantages of economic growth but at the same time to profit from the advantages. The extension agent has the responsibility to stimulate such an unpleasant choice so that it becomes a conscious decision. There may be considerable difference of opinion between those involved about which value judgement must be used. People sometimes are anxious about expressing difference of opinion because this may cause a conflict which leads ultimately to a political choice. However, if these differences are expressed there is a greater probability that a well considered decision will be taken. People often admire others who express different opinions convincingly and honestly, or who defend openly different interests from those they hold themselves.

In addition to value judgements, insight into the causes of a certain undesirable process or situation play a role in determining the desired situation. Are declining cassava yields caused by disease, by exhausting soil fertility or by soil erosion? Naturally it is important to have expert sources of this type of information when defining extension goals, but the expert can be a successful farmer.

The problem with determining which situation we wish to achieve is that we no longer believe we can predict what the future will be. For example, it is generally believed that biotechnological research will have an important influence on agricultural development, but nobody knows what form this influence will take. We can base our planning of extension programmes only to a limited extent on developments we expect in the future. The extension organization must be capable of adjusting rapidly to developments as soon as their implications can be seen clearly. Management should strive towards flexibility, towards keeping as many options open as possible. This is why it is important to have the capability of deploying capable extension agents as soon as a new problem arises.

Choice of objectives for an extension organization will be influenced by the opportunities its management can see for achieving the desired situation, as well as the vision management has of the present and the desired situation.

We must ask why the desired situation has not been achieved. Is there insufficient knowledge or skills? If yes, which knowledge and skills are lacking among which groups of farmers? Are there some undesirable habits or characteristics that have not been identified? Does social control make it difficult to achieve the desired changes? Does the social structure of the society cause or maintain an undesirable situation? Are there laws which lead to undesirable consequences? If so, which laws, and what are the consequences? Questions like these must be answered very precisely if concrete goals are to be formulated for the extension programme. Extension agents require concrete goals if they are to be given clear directions how to behave.

We must direct our efforts at variables which can be changed by extension and for which we have the manpower and resources available to bring about change. Unfavourable weather conditions can threaten farmers' livelihood, but, to give an extreme example, it is senseless to make changing the weather an extension goal. An extension agent is more likely to be successful if he or she tries to make farmers aware of the advantages of constructing a dam for holding irrigation water.

During our discussions of the decision-making process we have pointed out that we must adapt our level of aspirations to what is possible to achieve, especially if it appears that we cannot achieve what we want with the resources available. Similarly, we may have to adapt an extension programme to what is possible. Achievements possible with fixed quantities of manpower and financial resources will depend on farmers' wish for change and their capacity to implement these changes.

We have shown already that extension can achieve limited changes. More can be achieved with an integrated policy of which extension is a part. The significant increase in Indonesian rice yields in recent years has been obtained by an integrated development policy which included agricultural research, agricultural extension, input distribution, marketing and price policy.

Some of the most important choices that must be made follow from Chapter 2:

- Does it concern changing individuals, their society or both?
- Do we strive for changes in a certain direction, such as higher yields, or do we want to improve the decision-making ability of farmers?

This is related to whether we want:

- to provide the farmers with a solution to their problems;
- to help them find their own solution; or
- to teach them how they can solve similar problems in the future.

Hierarchy of goals

Many extension programme goals are expressed vaguely, for example, helping farmers to help themselves, which is not very helpful. Changes in knowledge, attitude and behaviour sought by the extension agents must be specified exactly if they are to deduce which educational experiences must be included to achieve these goals. We must differentiate between different levels of goals in order to specify programme goals concretely. This means the extension agents generally will have to develop a hierarchy of goals. At the top is the ultimate goal which they strive for because of its intrinsic value. The only answer to the question why they strive for this goal is: 'Because they want to'. This ultimate goal often is determined when setting up the extension organization. Programming is directed towards a number of intermediate goals which show how we think we will achieve the ultimate goal.

Thus the extension agent expects that achievement of the lower level intermediate goals will contribute to reaching the higher goal. Hence an intermediate goal is a way of achieving higher goals, leading to a chain of goals. Several intermediate goals each will contribute to the realization of the same higher goal. The example in Table 7.1 may make this clearer.

Table 7.1 Example of hierarchy of goals

Ultimate goal:	improved welfare for the citizens of Country A.
Intermediate goal 1:	higher income
	(a) farm
	(b) non-farm
Intermediate goal 2:	higher productivity on farms
Intermediate goal 3:	higher milk yields per cow + many others
Intermediate goal 4:	(a) breeding for increased genetic potential
	(b) improved animal nutrition
	(c) improved health care
	(d) improved housing
	(e) improved milking techniques
Intermediate goal 5:	(a) farmers who know the optimal way to feed their cows
	(b) impoved feed production on the farm
	(c) local availability of supplementary feed

The extension organization can work out a number of concrete activities from a scheme such as this. For increasing farmers' knowledge of cattle feeding they could:

(1) organize a team of research workers, extension agents and interested farmers who develop recommendations for cattle feeding in the local situation;
(2) write a series of articles for the local newspaper which later is distributed as a leaflet;
(3) give a lecture to the local farmers' organization on cattle feeding;
(4) organize a course on cattle feeding for the farm women who are responsible for this task on farms in this district;
(5) train the home economics extension agents in cattle feeding so that they can provide follow-up to the course mentioned above;
(6) organize a competition in cattle feeding.

An extension agent who analyses various goals in this way can see whether or not lower-level goals really contribute to achievement of those at a higher level, and if perhaps there are intermediate goals that will do this better. Naturally the extension agent cannot decide what a farmer's ultimate goal must be. However, the extension agent can contribute to the farmer's choice of intermediate goals because, by means of professional expertise, he or she often can see which of the intermediate goals is the best for achieving the ultimate goal. For each intermediate goal the extension agent can choose (an) extension method(s) which is/are most suitable for achieving this goal.

It is also meaningful to differentiate between long- and short-term goals in this way, as well as between goals on a national and local level. These latter two levels must be in tune with each other.

We have pointed out the importance of formulating goals precisely, if possible numerically. A goal of 'improving cattle feeding' is not precise enough. Success or otherwise of the problem could be assessed if the goal was expressed as: 'within two years the major recommendations should be known to and accepted as desirable by half of the farm women who have charge of feeding at least two dairy cows'. The goals should indicate what the farmers are expected to know, to believe and/or to do as a result of the extension programme.

Sometimes programmes specify goals for the activities to be undertaken by an extension agent. For example, placement of an article in the local newspaper each month. This is not a goal but only a means.

It is not unusual for an extension organization to try to solve simultaneously all problems identified in a problem analysis. Such attempts have little chance of success. It is much more effective to concentrate attention on one or two problems, thus identifying issues for the programme to focus on. Problems should be selected because they are important, there is a chance of solving them and the target group is interested in extension on these topics. When the problem has declined in importance because of extension activity, the extension organization can refocus attention on another issue. Even partial solution of the first problem will increase the target group's trust in the extension organization. Hence the shortest path for the extension organization to follow from the present to the desired situation is not always a straight line. The desired situation may be reached faster if extension is given first about problems the farmers regard as important. If the extension programme goal is to improve farmers' irrigation techniques, but the farmers are seriously worried about an insect plague in their crop, it will be necessary first to discuss what they can do about this plague. By starting to talk about irrigation techniques the extension agent would indicate to the farmers that he or she is not really interested in their problems, and hence they would not be interested in the message. Their attitude will be quite different after the extension agent has helped to solve their insect problem.

We can ask the following questions about goals when evaluating an extension programme:

(1) Are the goals clearly in tune with the farmers' problems? Have they participated in choosing the goals? Is our organization more likely to achieve these goals than another actor in the AKIS?
(2) Are the goals spelled out clearly? Has an outsider read and understood what the extension agents are trying to achieve?
(3) Are the goals important both in the eyes of the target group, the extension agents and their superiors?
(4) Do these goals contribute optimally to the implementation of a larger change programme of which the extension programme is a part?
(5) Do the goals specify clearly what the farmers will know, want, can and/or do, or do they indicate only what the extension agents will do?

(6) Given their skills and the means available, can the target group achieve the goals?
(7) Are the goals specified in such a way that clear conclusions can be drawn for choosing the extension message and methods?
(8) Are the goals specified so that we can evaluate whether or not they have been achieved?

7.3.2 Selecting the target group

A good extension programme is directed at a precisely defined target group because clear decisions can be made about choice of goals, of content, of methods and resources and of the manpower needed to achieve these goals. For example, we can choose to teach farmers how to use fertilizers. There is a higher probability that we will know what the extension agents should do if we specify our target group as 'those farmers who grow upland rice on sandy loam soil and who have not used fertilizers', rather than simply as 'farmers'.

The target group normally is larger than the group of farmers the extension agent contacts personally or via the mass media. We also hope to contact others indirectly through those who have direct contact with the extension agent. We may attempt to do this consciously by choosing an intermediate target group such as opinion leaders (which we discuss in Sections 5.6.4 and 10.9). We also can look for other groups which may have considerable influence on the ultimate target group and are prepared to work jointly with us. Teachers often are selected for this role, but they are not prepared to cooperate with all extension programmes because they too must choose their own priorities.

We must also ask ourselves whether we should direct our attentions to those who have asked for extension help or to those who know little of the results of scientific research and/or do not take this into account in their behaviour. This has been discussed in Section 5.6.5.

It is usually desirable to segment the target group into several different groups which can be approached in different ways. Thus in educating children we teach first form children differently from those in sixth form because of differences in previous knowledge and interest. We do not always make such a distinction in extension programmes, so that the sixth form pupils, who are more vocal, receive more attention than the first formers.

With this target group segmentation we can try to form sub-groups which have to make more or less the same decisions, which need the same type of help with their decisions, or which can be reached through the same channels. For example, for plant protection we could segment farmers who grow different crops, or those who spray their crops themselves, who hire a commercial sprayer, the sprayers themselves, pesticide distributors and politicians and civil servants who make and control regulations regarding plant protection. We often have to segment the target group according to their available resources and their access to inputs, credit, etc. because these resources have a major influence on the recommendations they can follow.

Different target groups will have different information needs for their decision-making. The extension agent must take into account that information intended for one group may reach another, perhaps causing unwanted effects. It is important to analyse differences in information needs and access between men and women. We discuss this issue in Section 10.6 while examining the roles of male and female agents.

Policy makers also can be an important target group because they have to be well informed about the situation, the problems and farmers' reactions if they are to develop effective policies.

We can ask the following questions when selecting the target group:

(1) Is it clear on which group the extension agents will or will not concentrate their attention?
(2) Have the reasons why they should concentrate on this group been well thought out? Do we know enough about the target group's specific extension needs and the way we can reach the group?
(3) Can we achieve our goal if we succeed in reaching this group?
(4) Have we segmented the target group? Is each segment more or less homogeneous with regard to extension needs and/or the way the segment can be reached?
(5) Is it possible to reach the target group with the means and manpower available?

7.3.3 Selecting extension content

The contents of the extension message depend very much on the goal and the target group, and also on the extension strategy. It is important that the message, which we might also call the recommendations, can be implemented by farmers using resources they have already, and with inputs available locally. For those farmers with meagre resources who can only bear very small risks, we can start with recommendations which require no extra expenditure, but which will increase yields or reduce costs. Examples of these might include seeding at the right depth or spraying pesticides at the right time. Farmers may save some money by applying these recommendations which they could invest in inputs known to increase their yields.

We often assume a message is ready for use in extension before it is in a suitable form for sending to farmers. For example, many research results may appear to be relevant, but first they must be integrated creatively into a practical usable solution. Innovative farmers and extension agents jointly play roles in creating such solutions. A solution must first be tried on a limited scale so that any difficulties can be sorted out before it is used on a large scale. Thus, test model cars are produced and tried before a new model is released onto the open market. During the early stages of production there may still be deficiencies to correct. Extension agents who offer farmers innovative solutions should follow a similar procedure, as a recommendation which fails may seriously undermine trust in them. Solutions to agricultural problems often are location-specific. They can be developed through field testing at various trial sites and by learning from farmers' experience.

Sometimes the knowledge required to solve a problem is not available. Extension agents then will have to give extension advice to research workers about farmers' problems in order to stimulate them to develop this knowledge, or develop this knowledge themselves jointly with the farmers.

Methodological considerations also play a part in choice of the message contents. In the first place, questions from the target group as well as the need they feel for information are of great importance. Furthermore, the contents should relate closely to the knowledge and attitudes of the target group if they want to use this information for decision-making and formation of opinions. Extension agents with specific technical expertise often assume their target group members know more than they really do. Hence the extension message is not understood. In a number of countries we have noticed the following weaknesses in the recommendations extension agents give to their farmers:

- Farmers will increase their yields, but decrease their income by following these recommendations.
- Farmers are not advised where they should invest the limited amount of capital they have in order to get the highest rate of return, but extension agents assume implicitly that farmers have an unlimited amount of capital.
- Advice given for one crop will increase the return from that crop, but decrease the return from other crops, because the available labour and capital is used for the first crop.
- The amount of risk farmers are able and willing to take is not discussed.
- The profitability of investments is calculated on the basis of the official interest rates.
- Social status is not recognized as a legitimate goal of a farmer.

We will not discuss these kinds of problems in detail, because they are discussed in books on farm management and farming systems research, although all too often extension agents' training in these subjects is weak.

We can judge whether the contents of an extension programme have been well chosen with the help of the following questions:

(1) Do the contents match the chosen goals?
(2) Are the contents described precisely?

Are they based on the latest scientific insights and on the experiences of successful farmers?
Have they been discussed in detail with a specialist in this area?

(3) Are the contents adapted to the time available?
Are the farmers overloaded with new information?

(4) Do the contents link up well with what farmers already know, are able, are willing to do and actually do?

7.3.4 Selecting extension methods

The choice of extension methods has been discussed at length in Chapter 6. There we saw that it is usually most effective to use a combination of extension methods. The preferred methods depend on:

- the goal;
- the size and the educational level of the target group;
- the level of trust between the target group and the extension agents;
- the extension agents' skills; and
- the manpower and resources available

It is important also to tie in with the ways in which the target group usually acquires its new information.

Cost is a prime consideration when selecting the method. Advertisers pay much attention to this point, taking as their criterion the cost per target group member effectively reached. Extension has paid less attention to costs, partly because extension agents' cost consciousness is less developed than in the commercial world, and partly because it is difficult to find a good criterion since different methods have different effects. We must account not only for the money spent but also for the time given by the extension agents. The latter usually is the most expensive part of an extension programme.

The major goal in the early years of an agricultural extension organization often is to increase crop yields and animal production. After some time more attention is paid to improving production efficiency. Such a change in goals also requires a change in methods. The crop and animal production extension agent often, but not always, can make clear recommendations based on research results. Farm management analysis is more a method for making good decisions, but only the farmers can make these decisions because they depend on values. For example, they alone know whether or not they will be able to sleep at night if they take the risk of using credit to expand their enterprise. Hence this change in goals requires a more non-directive approach. There have been problems in some countries in which the extension agents received in-service training in farm management analysis but were not taught how to use non-directive methods. They sometimes made recommendations about management of the farm where these were not justified.

The effectiveness of the programme is influenced not only by the choice of methods but also by the way in which these methods are used. We can judge whether or not the method is well chosen from the following questions:

(1) Is the chosen method adapted to the question of whether we wish to achieve a change in knowledge, skill, attitude or behaviour?
(2) Are the educational activities clearly specified so that we know what the farmer will see, hear, discuss and carry out?
(3) Are the different methods integrated in such a way that they reinforce each other?
(4) Does the planned scale make it possible to carry out all these activities well?
(5) When choosing learning activities, have the needs, skills and means of the target group been considered adequately?

7.3.5 Organization of activities

We can differentiate between organization of activities within the extension programme and organization of cooperation between contributors to the extension programme and/or a wider change programme of which the extension programme is a part. A well run extension programme requires definite commitments about who will contribute what and when. For this reason an extension programme should be put on paper so that all agreements are recorded. For example, it could be a condition for a well run programme that extension agents acquire competence in a new topic. They then must agree who is responsible for their training and how that person will train them. In fact, we often need two related programmes: an extension programme to achieve change among farmers and a training programme to enable extension agents to perform well the new tasks they are allotted in the extension programme. Usually this should be a programme of training in what to teach as well as how to teach, because a change in content often requires a change in methods. We return to this training programme in Section 10.4.

At the same time the extension programme should not be too rigid. Extension agents should be flexible in their reactions to changes in the situation, and especially to the reactions and emerging needs of the target group. Agreements about who should do what and when only should be made in so far as the situation is clear. One can go further by reaching agreement on procedures which will be followed to make new decisions to adjust to the changing situation. In addition, the most likely contributors to the programme could put time aside in order to make their contributions. Also, the most likely developments can be assumed when planning, providing the plans are adapted regularly to new information.

We have shown already in Sections 2.2 and 7.3.1 that an extension programme is especially effective as part of a wider programme of change.

Hence the importance of ensuring the extension programme is in tune with these other aspects of the change programme. The coordination of programmes carried out by different government departments often is difficult. This coordination should take place at the project and provincial as well as the national level. This is only possible if the lower levels have sufficient autonomy or freedom of control from the national level to undertake the activities required for their region.

Extension messages should be given at the moment farmers need this information. Thus, information about the best variety to sow should be given just before it is time for them to order their seed. This may involve preparing extension materials and training extension staff on this topic well in advance. The extension programme could include a calendar which indicates when which messages are given and when the preparation and training should start. Messages about plant protection often are given when a specific crisis arises due to a sudden attack by insects or plant diseases. Most or all of the extension materials for such crises should be prepared in advance. It is usually known which kind of insects and diseases are likely to cause severe problems at some time. In much the same way, a fire brigade uses the time when there are no fires to prepare equipment and carry out training to ensure they can fight fires effectively when called upon. Different extension methods should reinforce each other through careful timing. Thus mass media can be used to create interest in demonstrations, and leaflets and pamphlets should serve as a follow-up to group discussions and demonstrations.

We can judge whether an extension programme is well organized or not by asking the following questions:

(1) Is there an action plan in which the time scale and responsibilities are clearly indicated?
(2) Do all concerned know what their tasks are and when they must carry them out?
(3) Are all the activities well integrated with each other?
(4) Do those concerned have sufficient time to prepare themselves for their tasks?
(5) Can the programme be carried out in the time agreed? Is it flexible enough to allow changes if necessary?
(6) Are all the necessary written and visual aids available or will they be ready in time? Who is responsible for making these aids?
(7) Does the programme need temporary experts, administrative or technical assistance? Who will see that these are arranged?
(8) Have other organizations or societies organized other activities for the same target group? Is there coordination between these activities and the extension programme?
(9) Have the formal and informal leaders of the target group been sufficiently involved with planning and execution of the programme?

7.4 The planning process

It can be seen from the above that decisions about goals, target group, messages, methods and organization influence each other. But it is impossible to pay attention to all the decisions at the same time. How can the extension agent overcome this problem? We think the ideas from Bos' decision-making model give the best lead (see Section 5.3). The extension agent who follows this method first has to make a series of general decisions about these points and then returns to them several times to decide each point precisely. Hence it is desirable to begin with the point that has the greatest influence on the other decisions. In general this will be the goal or the target group.

Thus a spiral model similar to that shown in Figure 7.1, emerges from the process of drawing up an extension programme. The extension agent can follow the spiral to a greater or lesser degree depending on time available and the importance of the extension programme. Naturally he or she would spend less time preparing for a talk than for planning the work of hundreds of extension agents for the coming five years.

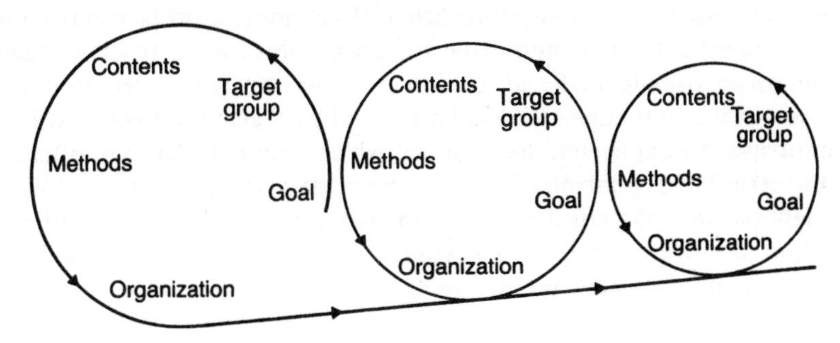

Figure 7.1 The spiral model of the process of planning an extension programme

As stated earlier in this chapter, we have made an artificial division between the pathway towards knowledge and the pathway towards choice. When following the pathway towards choice as described in Section 7.3, the extension agent simultaneously will have to pay attention to the pathway towards knowledge, as described in Section 7.2. The more exactly the goal and target group are determined the more the extension agent can specify what additional information he or she needs to collect in order to make the right decision on the pathway towards choice.

When discussing education processes we noted the important role played in this process by the desire for change. The first goal of extension agents must be to make farmers aware of the need to change if they do not already have such a desire. The change programme itself will succeed only when they have achieved this. Many extension programmes fail because the extension agents assume at the outset that farmers wish to change, whereas the farmers may be perfectly content with their present opinions and behaviour.

Extension agents not only have to stimulate farmers to form opinions and make decisions, but also often will have to help them implement the decisions. Many new problems arise, for example, when a farmer decides to combat increases in wages by keeping 60 milking cows instead of 20.

The extension agents have to give their full attention to such questions in order to show how they think they will solve the problems which arise in their programmes from putting such a decision into practice.

The decision about who contributes what to the process of planning an extension programme should be unanimous. It is particularly important for the specialists and generalists to integrate their contributions, and to clarify the role farmers or their representatives play in this process.

In Chapter 9 we discuss the way farmers can contribute to the planning process, and in Chapter 10 the teamwork between specialists and generalists when we consider the management of extension organizations. Our discussion of leadership also is important when considering cooperation between different levels of the extension organizations. Generally it seems undesirable to us to take an extreme viewpoint such as 'Everything must be run from the top' or 'Everything must stem from the grass roots'. An extension organization which tries to make all decisions at the top will never produce a programme that is in tune with local needs and local extension agents will not be motivated to implement decisions on which they have had no influence. On the other hand, some decisions must be made at the top to avoid division of attention, to make effective use of the mass media and to give sufficient attention to new problems such as energy conservation which research and policy developments have pointed to as important in the future.

7.5 Research needs

Perhaps the least research in extension education has been done on the planning of extension programmes (3). In many respects this leaves much of what we have written here as unsatisfactory. For example, the opinion we have defended that farmers' participation in extension programme planning increases the effectiveness of these programmes is more a belief than a well established fact. Amongst other things, we need:

(1) a description of the way an extension programme is planned and the role different staff members of the extension organization, farmers' representatives and cooperating organizations play in this process;
(2) analyses of the extent to which these people agree about the extension organization goals, the best ways of achieving these goals and the role that each of them should play in formulating the extension programme;

(3) a comparison of the programme plan and actual implementation;
(4) to understand the effects achieved with combinations of different extension methods;
(5) to know the extent to which there is cooperation between the extension organization and the many other services and voluntary organizations, and how this evolved; and
(6) to know how the AKIS functions.

7.6 Chapter summary

Programming extension work helps to ensure that:

- attention is concentrated on farmers' problems that are likely to be important to them in the future, as well as to the country, and which also interest them;
- attention will be concentrated on those target groups which are most important for achieving programme goals;
- the most effective combinations of extension methods are used;
- the activities of extension agents and others are integrated and organized as effectively as possible; and
- extension agents' training is directed at the major changes in their tasks.

Achievement of these requires decisions about the goals of the extension programme, the target groups that are to be reached, the contents of extension messages, the combinations of extension methods and the organization of all contributions to programme implementation. As these decisions mutually influence each other they can best be taken in a series of steps. Rough decisions are made during the first step and progressively refined.

Programme goals should be based on careful analysis of the target group, their present situation, their desired situation for the future, the reasons why the desired situation has not yet been achieved and the possibilities for achieving it. Other policy instruments may be used in addition to extension and must be attuned to each other.

Clear decisions should be made regarding the contributions different extension agents, farmers' representatives and the representatives of other organizations involved in agricultural development will make to the process of planning the extension programme.

We consider planning of extension programmes is the part of extension education most in need of further research. Table 7.2 reviews the questions that an extension programme tries to answer.

Table 7.2 Main points of an extension programme

(1) Initial situation
  (a) What is the initial situation regarding problems which this organization can include in its programme in the region where extension will be offered?

(2) Formulating the goal
  (a) What do farmers, the extension agents, the government and other relevant organizations see as the desired situation?
  (b) Why is this (desired) situation not yet achieved?
  (c) What must the farmers:
    - know (knowledge)
    - want (attitude)
    - be capable of (skills)
    - do (behaviour)?
  (d) How can the extension programme contribute to the achievement of the agricultural development plan?

(3) Target groups
  (a) Which groups of farmers must be reached?
  (b) Are there important differences within these groups?
  (c) In other words, must all farmers be approached in the same way or is segmentation desirable?

(4) Contents
  (a) Which aspects of the message contents will be emphasized?
  (b) Which educational experiences can be used to help farmers achieve the goal?
  (c) What must the farmers see, hear or do?

(5) Methods
  (a) Via which extension methods can these learning experiences be achieved?
  (b) Which communication channels will be used for this?

(6) Organization
  (a) Who is responsible for organization of this programme?
  (b) Who will participate in planning this programme?
  (c) Who shall undertake which activity and when?
  (d) What budget, how much manpower and other resources are needed to implement the programme?
  (e) How are the required resources obtained?

(7) Staff training
  (a) Which skills do different categories of extension agents need to implement the programme?
  (b) To what extent do they have these skills?
  (c) Which training programmes are needed to increase their skills? (See Section 10.4.)

(8) Evaluation
  (a) How can we determine the extent to which the desired programme results have been achieved?
  (b) Through which process have these results been achieved?
  (c) How will this information be used for improving extension?

☞ Discussion questions

1 Which process would you follow to decide on the extension programme goals for an area in your country with 15 Village Extension Workers?
2 Suppose it has been decided to start an extension programme on growing fish in fish ponds in your province. Which information about the target group should be collected so that you can plan an effective programme?
3 Outline the process you would follow to decide which contributions on plant protection in maize cultivation in your country would be made by which staff member of an extension organization?
4 An extension programme chooses poor farmers with a low level of education who seldom seek extension help as a target group. What can extension agents do to reach this group?
5 Discuss how principles of programme planning can be applied to training extension agents.

Guide to further reading

Almost every textbook about extension contains a chapter on programming. Up-to-date is:

Forest, L.B. and Baker H.R. (1994) The Program Planning Process. Chapter 10 in: *Extension Handbook: Processes and practices* (ed. D.J. Blackburn). Thompson Educational Publishing, Toronto.

Adhikarya, R. (1994) *Strategic Extension Campaign: A Participatory-oriented Method of Agricultural Extensions*. FAO, Rome.

A more detailed treatment is given in:

Sartorius, R. (1996) The third generation logical framework approach: Dynamic management for agricultural research projects. European Journal for Agricultural Education and Extension, 2 (5), 49–62.

Havelock, R.G. (1973) *The Change Agent's Guide to Innovation in Education*. Educational Technology Publications, Englewood Cliffs.

Kotler, P. (1972) The elements of social action. In: *Creating Social Change* (eds G. Zaltman, P. Kotler and I. Kaufman), pp. 173–85. Holt Rinehart and Winston, New York.

8 Evaluation and Monitoring

'They who are satisfied with their work have reason to be dissatisfied with their own satisfaction.'
Multatuli

8.1 Introduction

We use evaluation to determine whether an extension programme has achieved its goals and whether these goals could have been achieved more effectively in a different way. It enables extension administrators, managers and agents to learn more effectively from their experience by systematic observation and analysis of this experience. It complements the data gathering for planning the extension programme with the collection and analysis of data for assessing the extent to which the programme has achieved the desired changes. In doing so it gathers much of the data required for planning the next extension programme or to improve the present programme.

Evaluation is an action-oriented management instrument and process. Information gathered in this process is analysed so that the relevance, effect and consequences of activities are determined as systematically and objectively as possible. It is used to improve present and future activities such as planning, programming, decision-making and programme implementation to achieve extension policy goals more effectively. It includes value judgements of the extent to which these activities have proved to be worthwhile in comparison with the resources used.

Some evaluation will be done by researchers applying social science methods. However, most evaluation will be done by extension agents for whom simpler appropriate and less time-consuming methodologies have to be developed. The agents often will restrict themselves to a systematic analysis of the observations they make in the normal course of their work, although sometimes they will be able to collect additional information by questionnaire.

Extension agents may use evaluation in different ways. The general intention is not to give them a pat on the back if they have done well, or to punish them if they have done badly. However, it is hoped they will be motivated to do their work better if they are given more insight into the results of their work. The most important result often is improvement of the programme that has been evaluated. Evaluation information helps agents to set more realistic goals and assists them to discover more effective methods to reach these goals. Evaluation plays the same role in an extension programme as feedback in the communication process. Extension agents operate in the dark without evaluation information and do not know if they are still on the right track. There is also much interest in evaluation because it may provide data to convince politicians and administrators that the budget for the extension organization should be enlarged or at least maintained at current levels.

Evaluation may have several functions for extension agents. It can provide information on which extension agents may base their decisions, although these decisions usually are based more on an image of reality built up from many sources of information rather than from one alone (compare with Sections 2.4 and 7.2). Evaluation can add something to this information base, thus causing some gradual change in the image of reality. For example, if evaluation research shows that coconut growers make less use of extension advice than crop growers, the extension agents can think about whether this is desirable, whether it is possible to change the extension programme to provide more help to the coconut growers and any other effects they could expect from such a change.

We have to be clear about what we hope to achieve before we start to evaluate an extension programme. Here we focus on evaluation for increasing the effectiveness of the extension service. First we must answer the following questions:

(1) Which decisions must extension agents take?
   (a) How can we improve the extension programme?
   (b) Should we continue with the programme, enlarge it or reduce it?
   (c) How can we justify the money we spend on the extension programme?
(2) What information is required to decrease as much as possible the uncertainty in making these decisions? Who needs this information?
(3) Can we expect to obtain some of this information through evaluation?
(4) Will the benefits to better decision-making gained from the evaluation and/or monitoring exceed the costs?

If we do not answer these questions there is a high probability that evaluation and/or monitoring would produce an interesting report for extension agents but would have little effect on the improvement of their work. It is very difficult to decide which combination of extension methods will work best when setting up a new extension programme. We can begin on a small scale with a pilot project, and evaluate this thoroughly so that work can be more effective when starting a large scale programme. Interim evaluation of programmes is most important because the information obtained helps us to adjust current extension programmes.

Evaluation of a completed programme provides no useful information for improving that programme, but the experience gained in this way can be valuable for planning future programmes.

Under current economic conditions governments increasingly require evidence that the money they spend on agricultural extension is well spent. Evaluation can provide some of the information required for this accountability. The top level managers of the extension organization usually are more interested in this information than the field extension agents who might have to collect the data.

Farmers in industrialized countries can receive information from many different extension organizations. They have to choose which organizations they will use, applying such criteria as reliability of the information, ease of obtaining it, adaptation of the information to their own situation and cost. An evaluator might test the quality of different extension organizations to help farmers with their choice in much the same way as consumers' unions test the quality of different products.

One of the problems with evaluation is that extension agents often see it as a threat, especially if they lack self confidence or are not convinced their superiors value their work. This can be a serious problem in cultures where criticism might cause loss of face and is not seen as a positive way to help agents improve their work. Good extension agents must not doubt the value of their work if they are to speak about it with conviction. On the other hand, an evaluator must follow a zero hypothesis that the extension work has had no effect, and then try to reject such a hypothesis. Feelings of being threatened make it difficult for an extension agent to work with the results of evaluation research. The evaluator can minimize this by giving the extension agents the impression that they are being assisted to do their work as well as possible rather than being checked on. Therefore, amongst other things, it is important that the extension agent helps to formulate the questions to be investigated, as well as assists with interpretation of the results.

It is usual to distinguish between evaluation and monitoring (Casley and Kumar, 1988). *Monitoring*, which is derived from the Latin word 'to warn', is seen as a management technique in which extension agents collect data on the way in which the extension programme is implemented and the problems it faces in trying to stay on the right track. In this way management is soon informed if the programme deviates significantly from the planned pathway, and is given reasons for such deviation. This allows management to take rapid steps to implement the original plans or to adjust them when they were not realistic. For example, let us suppose that every Village Extension Agent in a particular programme is expected to visit five farmers a day who are representative of farms in his area. Monitoring shows that the poor transport situation makes it impossible to visit more than two farmers a day in some parts of the country, while the social structure makes it necessary first to visit the village chiefs. Management then will have to take measures either to implement the original plans by providing better transport or to adjust the plans to the field situation.

Monitoring of this type is feasible when the extension programme is relatively simple and the extension agents have to deliver nearly the same message to all farmers. It is difficult to implement when extension agents are expected to help each farmer with his specific problems.

Most of our attention in this chapter will be directed at evaluation and evaluation research. We will discuss monitoring only where there are clear differences in research methods.

Evaluation of an extension campaign normally involves a review of results achieved in relation to the extension given, on the basis of certain established criteria. In this chapter we will discuss which criteria can be used. We then will outline how data required for the evaluation can be collected. Here there are some difficulties to distinguish between changes resulting from extension and those having other causes. Furthermore, many extension programme objectives are vaguely defined and tend to change frequently.

8.2 Levels and criteria for judging extension programmes

In evaluating extension programmes we try to judge the extent to which the goals have been reached and the right steps have been taken towards reaching these goals. In Section 7.3.1 we discussed how there is a hierarchy of goals. This implies that we can evaluate an extension programme at different levels. At an early stage we can evaluate how the programme has been planned, and some time after the programme has finished we can evaluate what the consequences have been for society. We also can evaluate at one of the intermediate levels which are reviewed in Table 8.1 based on Bennett (1).

The higher levels in Table 8.1. generally are the result of the lower levels. In principle evaluators can direct their attention to each of these levels. The higher the level becomes that they are observing, the more accurately they will be able to state how far the ultimate goals of the extension programme have been reached. At the same time, as a rule it becomes more difficult to find suitable measurement criteria for changes at higher levels. It is also more difficult to prove that changes at these levels are the result of extension activity and not of other factors. It is rather easier to demonstrate that this is the case if attention also is paid to the lower levels during evaluation, although the manpower and means may not always be available.

Table 8.1 Levels for judging an extension programme

Level	
8	Consequences for society.
7	Consequences for the target group.
6	Behavioural changes in the target group.
5	Changes in knowledge, attitude, skills, motivation and group norms.
4	The farmers' opinion about extension activities.
3	Farmers' participation in extension activities.
2	Implementation of the programme by extension agents.
1	Programming of the extension activities.

Suppose that the extension service of a nation advised their farmers about modern methods of growing rice, with the goal of increasing national income. If the income indeed has increased, we could say with increasing certainty that this was due to extension if we know that rice production has increased (level 7), that the farmers have used modern rice growing methods (level 6), that they know about the results of research into rice growing (level 5) and that they have visited demonstrations of these methods (level 3). The evaluator also can aim at a higher level and try to determine the consequences for society of increased national income.

Choice of the level is very important for the evaluation criteria used by the evaluator. We now will discuss some possible criteria for each level.

Level 1

We have given a large number of questions for extension programming in Chapter 7 which can be used to decide whether we can expect to achieve the desired results with a particular programme.

Level 2

This level usually is called 'monitoring' rather than 'evaluation'. The monitor can observe the manpower and resources required for effective implementation of the programme. Regrettably, little attention is usually paid to this aspect despite its importance when assessing the benefits achieved in relation to the costs of the programme. The evaluator also can monitor whether or not the programme has been carried out according to plan, and if not, why it deviates from the plan. Such deviations may be desirable, providing they rely on conscious decisions based on new information.

Finally, the evaluator often can derive unwritten goals from the way in which the programme has been carried out. It already has been pointed out in the introduction to Chapter 7 that extension programme goals are not always formulated clearly. If the evaluator sees how the extension agents operate he or she will ask 'Why do they work this way?'. Then the evaluator should put these suppositions about their goals to the extension agents and ask if they consider the suppositions to be correct. In this way the evaluator can help extension agents to define their goals clearly, which often is one of the most important ways to improve extension.

Level 3
The extent to which farmers participate in extension activities is an important evaluation criterion. The evaluator can check how many publications they read, what meetings and demonstrations they attend, what requests they make for extension advice and so on. It is important here to know not only how many but also which farmers participated in these activities. The evaluator then can compare this information with that about the target groups the extension agents are trying to reach. If these groups are not clearly defined the evaluator then can ask whether in fact they were trying to reach these or other farmers, thus contributing to a more conscious choice of target groups.

Level 4
Farmers' opinions about extension activities often provide important opportunities to make interim adjustments to the extension programme. The evaluator must watch both extension content and method. Is the extension programme adjusted to the needs of different categories of the extension agent's clients? Do they understand the extension agent clearly? Farmers' opinions on these matters will explain to a large extent their participation in extension activities.

Level 5
The evaluator usually will have to use questionnaires to measure knowledge, attitude, motivation and group norms, preferably on at least two different occasions. This makes measurement very expensive at such a level. However, it is essential for us to obtain facts about these changes if we are to gain insight into the success of extension as an educational activity.

Level 6
Adoption research has been directed especially at determining this behaviour change. We can draw useful conclusions for extension from this work if we know not only the extent to which methods recommended by extension agents are applied, but also who does or does not apply them and why. It makes a great difference if farmers who do not apply recommendations have not received extension advice, or if they are not convinced that the advice was correct, or if lack of resources prevents them from putting the advice into practice. Behavioural changes such as these usually are influenced by other factors in addition to communication with an extension agent.

Level 7
We can consider the hierarchy of goals shown in Table 7.1 for analysing the

consequences of behavioural changes achieved by the target group. Extension often will be directed at an intermediate goal. It is then important to see if it contributes to achieving the intermediate goal of the higher level and to the ultimate goal. For example, if people are inoculated against cholera because of an extension campaign are they necessarily healthier? When interpreting these consequences we must take into account not only what extension has achieved but also the effects of a larger development programme of which extension is a part. Thus, in the example above it is possible that extension convinced the target group that inoculation was desirable, but the inoculation programme itself was poorly organized.

If the consequences of farmers' behavioural changes are as expected, it indicates that the suppositions are correct about the hierarchy of goals on which the extension programme was based. Often we find that some categories of farmers change and others do not. We should analyse the causes and consequences if this is the case. In the rice cultivation example mentioned above, we need to know which farmers did not adopt the new rice technology, why they did not adopt it and what the consequences were for them that other farmers increased their rice production by adopting new rice production technologies.

Level 8

Changes in target group behaviour usually will have consequences for other groups in the population, and sometimes for the whole structure of society. For example, an effective family planning education programme may cause unemployment amongst teachers. We should know about such consequences in order to be able to judge whether or not expansion or contraction of extension in this area is desirable, or whether the goals or target groups should be changed.

We have pointed out already that opinions about whether the effects of extension are desirable or not ultimately rest on *value judgements*. The values of different people involved and the culture of the society in which they live influence which changes they consider to be desirable. For example, it is desirable that agricultural productivity is increased in a region, while at the same time some people are pushed out of agriculture and are unemployed? Some evaluators believe they must express an opinion on the basis of their own value judgements. However, they do not always say what these judgements are, and it is not always clear whether decision makers will have to reach a different conclusion if they do not share the same value judgements as the evaluators. Those who have to make decisions about extensions are not always interested in evaluators' judgements because they consider their own, those of farmers and of their superiors to be more important.

Nevertheless, it may be very fruitful if the evaluator stimulates discussion about the value judgements on which his extension work must be based.

These value judgements often are not discussed so as to avoid conflicts, although this makes it difficult to achieve consensus or to work harmoniously towards achieving certain goals. The evaluators can indicate the assumptions they make about relations between the extension advice given and the background value judgements. This can stimulate discussion between extension agents, farmers' representatives who participate in planning and the extension programme, and the extension agents' superiors about which value judgements were used as a starting point and which should be used.

Evaluation usually will show that extension has not fully achieved the initial goal. We will have only limited possibilities for improving extension if we know what the differences are between the initial goal and that achieved. We must also know why extension has achieved more or less than planned.

It is a human failing to look to others rather than to ourselves for the reasons why we failed to achieve our goals. Often it is the farmers who are considered to be lazy, conservative or stupid, although it may be more fruitful and realistic if the extension agents look for the causes in their own work methods. It may be that the target group is uneducated (which is a more correct term than stupid), but the extension agent should have taken this into account with good planning. Has planning for this aspect been correct? Has the extension agent really taken the target group's needs into account?

The extension agent is not always responsible for disappointing results. These also may be caused by factors beyond his or her control, such as unexpected price developments on the world market. Or collaborating organizations may not have contributed as much to the programme as promised. We then must find out why these promises have not been honoured so that we know what collaboration to expect when planning the next programme.

8.3 Collecting data required for evaluation

The purpose of the evaluation will determine which data have to be collected for evaluating an extension programme. We can distinguish between *formative* evaluation which gathers information for development of an effective extension programme, and *summative* evaluation which tries to measure the end results of a programme in order to decide whether or not it should be continued, expanded or diminished. Data collected may be quantitative or qualitative. The former are useful for measuring changes achieved as a result of the extension programme, while the latter provide information about the reasons why extension agents and farmers act in a certain way. There is more interest in using qualitative data in recent years because they have been most helpful for improving programmes.

8.3.1 Questionnaires

Most quantitative data are collected with a fixed questionnaire for interviewing a random sample of farmers and/or extension agents. Evaluators often realize at the end of the interview that they did not ask some of the most important questions. During the research process with qualitative data the evaluator tries to learn which questions should be directed to whom.

8.3.2 Experimental design

An experimental design in which comparisons are made between areas with and without the extension programme, or where additional extension activities have been undertaken in the experimental area, often is preferred to summative evaluation. For example, if the administrators of an extension organization wonder whether it might be useful to introduce the Training and Visit System (see Section 10.9) into their country they can first introduce the system in a number of randomly selected pilot areas and compare results obtained in these areas with changes obtained in randomly selected control areas. This research design allows us to separate the effects of the T and V System from other factors which influence agricultural development. Without such a control group it would be impossible to say whether the changes observed are the result of the T and V System or of any other factor, such as a change in product prices or a radio programme.

We also may work with randomly selected individuals rather than with randomly selected areas. For example, if we wish to know whether it pays to rewrite articles written by technical experts for a farm magazine using readability principles (see Section 6.1.2), we can rewrite one or two of these articles, print a version of the magazine with the original articles and another version with the rewritten articles, distribute both versions to a random selection of subscribers and measure the effects of both versions. It is important that the data are collected from a random sample of the target group and not, for example, only from those farmers who come to a meeting, because such farmers usually are more interested in extension than the average farmer. It is more important for drawing the correct conclusions to use the right sampling method than to use a large sample.

Most textbooks on evaluation express preference for such an experimental design because it increases the probability of drawing correct conclusions. However, in practice there are some difficulties with such experiments. A large number of variables influences the effects of extension programmes. Usually it is feasible to test only a few of these variables in an experiment. For example, if rewriting to increase readability we would use only simple words, a clear structure and focus on readers' problems or needs. Then it would be impossible to say which of these three factors is the most important in improving readability and understanding unless we are willing to use a very complicated experimental design. Normally we must restrict experimentation to some variables which are important from the viewpoint of theory or management. Usually the best we can do in the field is a quasi-experimental design because of the difficulties of implementing a full experimental design. For instance, it is almost impossible to achieve a random distribution of villages over control and experimental groups.

It can be a good practice to test an extension innovation first on a small scale before introducing it on a larger scale. Ideally the test situation should be similar to that of the large scale application, although this often is not the case. The pilot project may receive much more managerial support, perhaps from expatriate experts, than can be given after introducing the innovation to the country as a whole. Also, the fact that a baseline survey is required for an experiment means that the extension programme in the pilot area may be much more carefully planned than might be expected under normal conditions in which information about the situation, knowledge, attitudes and behaviour of farmers is more restricted. On the other hand a pilot project gives us the opportunity to learn the best way to apply an innovation in a local situation. For example, if the T and V System is first introduced to several districts in a country it is possible to learn how to adapt general principles developed in other countries to the local situation. It also gives programme staff an opportunity to learn how to work in a new way. Summative evaluation gives little information for this learning process.

In contrast with the physical sciences, there are hardly any laws which hold true under all circumstances in the social sciences. If we find experimentally that one extension method is more effective than another, our conclusion may no longer be true in a few years when the situation has changed (2).

8.3.3 Difficulties with experimental design

It can be useful working with randomly selected individuals if you wish to know how use of different mass media affects knowledge gain. Most extension programmes also have behavioural change as a goal. There are some difficulties with this design because behavioural change is highly influenced by interpersonal communication (see Section 5.6.4). Hence we prefer to experiment with randomly selected areas so that farmers in the experimental group are less likely to influence farmers in the control group. However, we may also encounter organizational problems with use of areas, for example, because of opposition from areas which do not receive the benefits of the supposedly better extension programme. Furthermore, the effect of an extension programme in an area depends to a large extent on the motivation and personal qualities of the local extension agent. This means that for a reliable design the extension agents also should be chosen at random, thus making the design very complicated, expensive and widely dispersed.

A good experiment requires the extension programme goals and methods to be kept constant. However, goals and methods tend to change constantly in a normal extension programme because extension agents learn from their experiences and modify the programme accordingly. Therefore, it is difficult for managers of the extension organization to accept the requirements of an experimental design because of changing situations, such as new research findings.

We also should evaluate whether or not the extension programme had unintended consequences. Usually you get some clue of such consequences during programme implementation. Hence, the baseline survey will not have measured variables related to these consequences. One such unintended consequence of some extension programmes has been a change in the division of labour between men and women.

We must take these difficulties with experimental design into account when planning an evaluation study. They do not mean that experiments are never possible, but that other designs must be considered. For example, you may wish to work with a time series to see whether the rate of change has increased after a new extension programme is introduced. In several countries yields were observed to increase in some districts following introduction of the T and V System. Despite the lack of careful comparisons with control areas, the reports of up to 50 per cent yield increases in a period of several years were enough to convince the director of agriculture to introduce this system on a wider scale.

Another research design involved comparisons of changes among those farmers who are reached by extension programmes with those who have no contact. Farmers in a true experiment would be assigned to the experimental and control groups at random, whereas in this natural experiment there is usually a select group which is reached by the extension programme. These farmers tend to have more education and resources and show more interest in change (see Section 5.6.2). Thus, differences between both groups will be caused partly by these selective factors and partly by the extension programme.

Simpler methods for gathering information can provide valuable leads to develop better extension methods and may have to be used if there are no funds to employ a research team. Currently many decisions about improving extension are based on impressions rather than on empirical data.

For summative evaluation often we will have to choose between:

(1) a carefully controlled experiment in which the effects of all possible extraneous factors are excluded, or
(2) an experiment in a natural setting in which extension is given in the normal way.

We can never be certain with the second way that differences we find between experimental and control conditions are due to the extension activity or to some extraneous factor. The disadvantage with the first way is that it can be difficult to generalize experimental results to the field situation.

Summative evaluation can contribute little to improving extension work in a programme that is implemented once only, unless clearly defined theoretical principles are tested in this campaign. Nevertheless, such an evaluation may be important for accountability of the organization. The results may determine whether it is easier or more difficult to obtain a favourable reaction to future requests for financial support. If the money was well spent by the organization on the present campaign, decision makers are likely to expect it will be used profitably in future for different campaigns.

8.3.4 Formative evaluation

Interest in formative evaluation has been growing in recent years. Evaluation increasingly is seen as a management tool for development of effective extension strategies. We may learn from an experimental design that a pilot extension programme failed to achieve expected results, but this design provides little information about why the pilot programme failed or what we could do to make the programme more effective. We require more information about how the programme actually was implemented, as well as opinions of extension agents and of different sections of the target group about the programme. Unfortunately there are evaluation studies in which, despite careful analysis of the effects of an extension programme, the inputs are not analysed simultaneously. It is impossible to judge whether the programme should be continued or not if we do not know how much it cost in terms of manpower, money and other resources. Similarly, one cannot suggest how an extension programme can be improved without knowing which extension methods have been used and how they have been used. It is essential in an evaluation study to compare the inputs with the outputs obtained. Pre-testing is one form of formative evaluation which can provide valuable information on how to improve extension materials. We can evaluate the experience gained from organizing a series of meetings in different villages and use this information to improve meetings in other villages.

8.3.5 Implementation of extension programmes

Extension programmes often are not implemented in exactly the way they were planned. There may be many difficulties associated with implementation such as transport problems, equipment breakdowns, delays in delivery of extension materials, or a reward system which stimulates extension agents to behave differently from the way they are supposed to behave. Extension management must be kept informed quickly about these problems so they can respond quickly, either by solving the problems as they arrive or by changing plans to make them more realistic in the light of existing limitations. Surveys of extension staff to discover their experiences with and their reactions to the programme can provide valuable information for improving it. The same is true with information collected in surveys of the target groups and users of the extension programme. We need to know the extent to which and the reasons why they read the publications, visit demonstrations, attend meetings and follow recommendations. We may receive more valuable information about these questions in an open discussion than from a mailed questionnaire, even if it comes from a smaller number of people.

8.3.6 Achieving the goal

In Section 7.3 we pointed out that the goal of the extension programme may be to achieve certain behavioural changes among the target group, such as an increase in their use of fertilizers. The goal also may be to help target group members achieve their own goals which will differ between different members. This makes summative evaluation difficult, but it may be useful to ask target group members about the extent to which they have participated in different aspects of the extension programme and how useful they believe this has been in helping them to achieve their goals.

8.3.7 Summative versus formative evaluation

The distinction between summative and formative evaluation is not always so clear as we have indicated so far. For instance, it is possible to measure changes obtained in a pilot programme, and at the same time to seek extension staff and target group reactions to this programme in order to obtain suggestions for improving the pilot programme. This clearly would reduce the value of the pilot programme for predicting the outcome of a full scale programme. The results will be different, and perhaps better, if the full scale programme deviates from the pilot programme. However, it can be a sensible way of making decisions about introducing a new extension programme.

8.3.8 Aspects to consider in the selection of data

The selection of data to be gathered will depend partly on theoretical considerations. How can the extension programme change farmers' behaviour, and which factors might influence this process of change? It also will depend on practical considerations. Which factors can the extension service influence? For example, it may be possible to influence the frequency with which farmers visit demonstration plots. Hence we need to know whether frequency of farmers' visits to demonstrations is related to their adoption of innovations. We cannot influence farmers' age so there is little point in asking questions about their age. We might base selection of questions to be asked on the criterion: What difference will it make to management decisions whether 20 per cent or 80 per cent of farmers answer this question affirmatively? As a general rule there is no point in asking this question if we cannot think of a difference beforehand.

Much information for evaluation will be gathered through interviews and questionnaires. Social scientists have paid considerable attention to development of these data gathering techniques. Their conclusions can be found in many textbooks on social science research methods and cannot be summarized here. However, some aspects require special attention for extension programme evaluation.

Interviews and questionnaires are designed with the intention of informing us about respondents' behaviour, knowledge and opinions. There is always the danger that respondents will reply in the way they believe they ought to act and think rather than in the way they actually act and think (3). This is a major problem with evaluation of extension programmes in which respondents have learned what they are expected to do or to think. Hence the wording of the questions and the behaviour of the interviews should convey the idea that it is in the respondents' best interests to provide information about what they actually think or do. This is the only way they can assist the extension agents to help them become as efficient as possible. Evaluation should avoid conveying the impression that it is an examination to see whether or not respondents have paid attention to their extension agents.

Open-ended questions to which the respondents can express their own opinions are especially useful for formative evaluation. Questions with pre-coded answers force respondents to answer within the evaluator's frame of thinking, whereas the most valuable information often comes from people with a different frame of thinking. For example, the evaluator may think farmers' acceptance of extension agents' ideas depends on the quality of communication between these two groups, whereas in fact the resources available to a category of farmers may limit their ability to accept these ideas.

The extent to which selection of respondents is based on a random selection of target groups is more important for the value of data collected than the sample size. The mail return questionnaire often is the cheapest way of obtaining information from a large sample, but replies to such a questionnaire usually are biased towards those with higher levels of education and greater interest in the extension programme.

Costs of data collection must be kept to an affordable level. One international aid donor insisted that a large number of vehicles be made available for the monitoring and evaluation unit in a country where extension had little impact, partly because no transport was available for extension staff.

It is usually the responsibility of field extension agents and their supervisors to evaluate the extension programme. Some extension organizations maintain an evaluation unit at their headquarters which has more competence in methodology and data processing. It may also be useful to involve outsiders such as university staff in the evaluation process because they may introduce valuable new ideas on extension strategies and methods.

With the trend towards more participation farmers are also involved increasingly with extension evaluation, not only for data collection, but also for deciding on criteria for successful extension programmes.

8.4 Chapter summary

Evaluation is used by extension organizations in order to learn from their experience in a systematic way.

Some of the questions that must be answered when planning extension evaluation are as follows:

(1) What is the extension programme goal? Is it to be derived from:
  (a) the extension programme itself;
  (b) discussion with extension agents; and/or
  (c) the activities of extension agents?
(2) What is the target group of the extension programme? Is it to be derived from:
  (a) the extension programme itself;
  (b) discussion with extension agents; and/or
  (c) the activities of extension agents?
(3) What is the evaluation goal? Will it help to decide on:
  (a) continuation of this programme;
  (b) improving this programme;
  (c) accounting for the money spent?
(4) How much manpower and resources are available for evaluation?
(5) What will the level of evaluation be and which criteria will be used to judge the extension programme?
(6) What research design will be used to evaluate the effects of the extension programme and the farmers' reactions to extension work?
(7) Which method will be used for gathering data on:
  (a) implementation of the programme;
  (b) effects achieved;
  (c) farmers' and extension agents' opinions of the programme?
(8) Who will gather these data? How will they be analysed, interpreted and reported?
(9) How will the evaluator cooperate with:
  (a) extension agents;
  (b) farmers?
    In which stage of the evaluation process?
    Which other methods will the evaluator use to promote the use of the results of the evaluation study?

Once an extension programme has been initiated, the process of monitoring it will involve asking the following questions:

- To what extent has the extension programme been implemented?
- What difficulties have been encountered in this implementation?

☞ Discussion questions

1 Indicate how an extension agent can evaluate a certain extension programme. You may use your own experience of advice to students on which course to take. Why do you select this level of evaluation?
2 Extension agents often consider evaluation by a research worker as a threat. What can an evaluator do to reduce this threat?
3 An extension programme teaches farmers to use better quality seed. How can an extension manager who has to supervise 100 VEAs evaluate this programme? Which additional assumptions do you have to make to be able to answer this question?
4 An expatriate evaluation expert notices that extension in a less industrialized country contributes to increasing average incomes, but at the same time it increases differences in income. On the basis of the value judgements which are generally accepted in his home country he is convinced that extension should be directed towards decreasing income differences. Should he try to promote extension activity that is in agreement with his own convictions? Why? If yes, in what way?

Guide to further reading

There are many books on evaluation and monitoring. The World Bank approach is discussed in:

Casley, D.J. and Kumar, K. (1988) *The Collection, Analysis and Use of Monitoring and Evaluation Data.* John Hopkins University Press, Baltimore.

Good practical books include:

Herman, J.L., Morris, L.L. and Fitz-Gibbon, C.T. (1987) *Evaluator's Handbook.* Sage, Newbury Park.

Patton, M.Q. (1997) *Utilization-focused Evaluation.* Sage, Thousand Oaks.

Raab, T.L., Swanson, B.E., Wenting, T.L. and Clark, C.D. (1987) *A Trainer's Guide to Evaluation: A Guide to Training Activity Improvement.* FAO, Rome.

Salmen, L.F. (1987) *Listen to the People; Participant–Observer Evaluation of Development Projects.* Oxford University Press, New York.

In evaluation we apply social science research methods. Good introductions in these methods for extension agents are:

Moris, J. and Copestake, J. (1993) *Qualitative Enquiry for Rural Development; A Review.* Intermediate Technology Publications, London.

Nichols, P. (1991) *Social Survey Methods: A Field Guide for Development Workers.* Oxfam, Oxford.

9 Participation of Farmers in Extension Programmes

Participation of the people involved in development programmes is often seen as a way to make these programmes more successful, especially for solving problems of poor people. This chapter tries to help the reader to analyse systematically what can and cannot be achieved by participation in his situation and how this participation can best be implemented.

We discuss different connotations of the word 'participation' and the way in which farmers and/or their representatives can cooperate with extension agents and others in planning and implementing the extension programme. We conclude by analysing the role NGOs play in rural development and extension.

9.1 What is participation?

'Participation' has quite different connotations for different people, such as:

(1) Cooperation of farmers in the execution of the extension programme by attending extension meetings, demonstrating new methods on their farms, asking their extension agent questions, etc.
(2) Organization of the implementation of extension activities by farmers' groups, such as meetings where an extension agent gives a lecture, organizing courses and demonstrations, publishing a farm paper in which extension agents and researchers write for farmers, etc.
(3) Providing information which is necessary for planning an effective extension programme.
(4) Farmers or their representatives participating in organization of the extension service, in decision-making on goals, target groups, messages and methods and in evaluation of activities.
(5) Farmers or their organizations paying all or part of the cost of the extension service.

(6) Supervision of extension agents by Board members of farmers' organizations which employ these agents.

We will pay most of our attention in this chapter to the fourth interpretation, 'participation of farmers in decision-making', but will also consider the second and third interpretations. Participation according to the fifth and sixth interpretations will be considered in discussing the work of NGOs and when we discuss the privatization of extension organizations in Section 10.8.

We see that participation through the inclusion of farmers can be a more efficient way to achieve the goals of the extension programme which has been formulated by politicians and extension officers. It can also be a goal in itself to give farmers more opportunities to influence their own future, as well as more power in the society (1). The danger of the first approach is clearly illustrated by a World Bank evaluation study which concluded:

> 'that agricultural projects including group participation often did not work because the groups were not committed to the project and acted more as an extension of the government than as organizations representing beneficiaries'. (2)

9.2 Why should farmers participate?

There are several reasons why it is desirable for farmers to participate in decisions regarding the extension programme:

(1) They have information which is crucial for planning a successful extension programme, including their goals, situation, knowledge, experiences with technologies and with extension and of the social structure of their society.
(2) They will be more motivated to cooperate in the extension programme if they share responsibility for it.
(3) In a democratic society it is generally accepted that the people involved have the right to participate in decisions about the goals they hope to achieve.
(4) Many agricultural development problems such as soil erosion control, achievement of a sustainable farming system and organization of a commercial approach to agriculture can no longer be solved through individual decision-making. Participation of the target group in collective decisions is required.

Participation makes it possible to achieve more profound changes in the ways people think. There will be less change in their way of thinking and acting, and these changes will be shorter lived if they follow recommendations of their extension agent obediently, than when responsibility for decisions is shared. (Compare the discussions of directive and non-directive group methods in Section 6.2.3.)

9.3 Who will participate?

Some Westerners have a rather idealistic view of participation which may not give sufficient attention to the fact that there are conflicts of interests in each society. An Indian farmer describes the consequences of these conflicts as follows:

> 'The rich have been sucking the blood of the poor. These tigers who are used to blood of people manage to get into all kinds of committees of projects and organizations in the garb of sheep. But as soon as they get the chance, they start sucking blood again, this time through the development organization.' (3)

We do not claim that participation of farmers in extension programmes always proceeds in this way. There are also situations in which the interests of leaders and followers coincide, such as to prevent the outbreak of a pest. There are also leaders who honestly try to serve the interests of their followers as well as they are able. However, even these leaders may fail to do so because they do not appreciate sufficiently that their own situation is quite different from that of the poorer farmers. In most societies it is mainly the more well-to-do and better educated male members of the group who are elected as leaders, for example, as headmen of the village. It is expected that they will also speak on behalf of the women and the other farmers on development issues.

If considering formation of a small group of ten women who wish to augment their income by growing vegetables for the market, it is possible to select members who are more or less equal in power and access to resources and who trust each other. This is not possible if selecting farmers to participate in planning an extension programme in a district with 100 000 farm families. Under these conditions one has to work with farmers' representatives. It then is important to specify how these representatives are chosen and how they are held accountable for their decisions to farmers they represent. They may be chosen through:

- farmers' organizations, at least in situations where a large proportion of the farmers belong to such an organization. The representatives can be held accountable for their actions at regular organizational meetings;
- elected local members of parliament or the district council. This may increase the influence of politics on extension which is not always desirable;
- extension agents who try to select people who are representative of the

whole target group. It may be difficult for these representatives to control whether the extension agents perform their job properly;
- village headmen and elders.

It is often claimed that participation makes it easier to reach the poorer and less educated target groups and the women. This is not always the case. It may be achieved in a small group in which all members participate in decision-making, although even in these groups it is possible that the poorer and less powerful members do not dare to disagree with a powerful member from whom they rent land, work as a labourer or borrow money. It does not necessarily mean that they accept the decisions taken as their own.

Participation of farmers can and often is achieved informally. The extension agents can listen very carefully to different types of farmers in their area in order to understand their needs, their goals and their opportunities. This information can and should play an important role in planning the extension programme. They can and should learn from the experiences of the most successful farmers and use this information in formulating extension messages which are tested in the local situation. They can discuss different opportunities for extension programmes informally with group members and adjust their plans on the basis of these farmers' ideas. Donors may not accept that this kind of participation meets their conditions for continued support for the project. However, in some situations with good village extension agents and extension managers who are eager to learn from their field staff, it may work more effectively than a formally organized participation system.

Many extension agents do not realize that they mainly have contact with the more innovative, better educated and larger farmers. The problems and opportunities of these farmers may be different from those with whom the extension service has little contact. It is therefore necessary to talk with a random sample of farmers for planning extension programmes. This can be done through Participatory Rural Appraisal. (See Section 2.4.3.)

9.4 Roles of farmers and extension agents in planning extension programmes

The roles which extension agents and farmers or their representatives can play in planning extension programmes depend on the one hand on the knowledge and competence of both groups. On the other hand they depend on what rights to make decisions each group is considered to have. One of the considerations in granting this right will be the expected impact this will have on the motivation of both groups to achieve the extension programme objectives. As a rule a good extension programme can only be developed by integrating knowledge and insights from both groups. A proposal for their division of labour is presented in Table 9.1. We have no evidence that this is

Table 9.1 Division of contributions by farmers' representatives and extension agents in planning an extension programme

	Farmers' representatives	Extension agents
1 Knowledge of the present situation	XXX	X
2 Knowledge of expected changes in situation	X	XXX
3 Knowledge of problems felt by farmers	XXX	X
4 Knowledge of possible solutions for these problems	X	XXX
5 Knowledge of the desired situation	XXX	X
6 Right to decide what is the desired situation	XXX	X
7 Right to decide what should be the target group	XX	XX
8 Knowledge of the consequence of the acceptance of extension recommendations by farmers	XX	XX
9 Knowledge of the effects of different extension methods	XX	XX
10 Knowledge of the reactions of farmers to previous extension activities	XXX	X
11 Knowledge of extension resources available	X	XXX
12 Knowledge of extension agents' interest and expertise		XXXX
13 Knowledge of effective procedures for planning a programme	X	XXX

The number of crosses indicates from whom we expect the greatest contribution.

the optimal division of labour, but we present it as a basis for discussion of how it might be aimed at in a given situation.

Farmers' representatives who participate in extension programmes cannot be expected to perform their new task very well at the outset. They must first be trained for it. Extension agents who have spent several months planning a new programme often are disappointed if the representatives do not make suggestions for improving it at the first meeting. The representatives can contribute to decision-making if they are involved with the most important choices to be made. The extension agents can discuss with them the different alternatives and their expected consequences, then jointly make the best choice. As the representatives become more experienced they can be expected to make their own alternative proposals, some of which the extension agents may not have thought about. For successful participation it is desirable that ordinary farmers as well as representatives gain insight into decisions made for the extension programme.

Reaching agreement about who will contribute what to the decision-making process is one of the difficulties with cooperation between extension agents and farmers' representatives when planning an extension programme. Hence it is recommended that the question of how everyone can best contribute to the programme be discussed explicitly. The framework given in Table 9.1 can serve as a basis for this discussion.

We must appreciate there are some disadvantages associated with leaving many decisions about the extension programme to farmers. One can be that many farmers will request services from extension agents rather than make decisions about fundamental changes in their own behaviour. It is often so difficult for us to change that we are not inclined to see the need for changes in our own behaviour. Farmers seldom ask for help with decisions about changes in their behaviour, despite the fact that such help could be important. For example, we think that the government agricultural extension service in the Netherlands took a long time to recognize that a portion of the farm population had to seek a job outside agriculture because of the influence of farmers' representatives.

Another disadvantage of participation referred to earlier is that extension agents have to divide their attention between many topics because each small group seeks advice about different problems.

The extension agent who recognizes these disadvantages and, if necessary, openly discusses them with the farmers participating in the planning process, will be able to make use of the advantages of participation without being hampered seriously by the disadvantages.

9.5 When to employ a participatory approach

Under what conditions is a participatory approach desirable and possible? In the discussion of directive and non-directive extension methods in Section 2.3 and in Chapter 6 we analysed when extension agents can tell farmers what they should do and when they should try to create a situation in which the farmers discover the answer for themselves. It will be clear that this discussion is related to the conditions under which a participatory approach to extension is desirable.

On a macro-scale Hayward came to similar conclusions (4). He states that a rather authoritarian approach can work well in situations where a new technology is available which will increase most farmers' income, while at the same time helping to achieve government agricultural policy goals, providing extension agents know the situation and the goals of their target group quite well. This has often been the case with the introduction of high yielding crop varieties in irrigated areas, as in many Asian countries where such varieties have prevented massive famines.

However, the situation is quite different in many rainfed areas, because there are large variations over relatively short distances in agro-ecological and socio-economic conditions, and often in the culture and the situation of the farmers. For example, extension agents often do not know in detail what access farmers have to markets. Under these conditions extension agents have to develop location-specific solutions for farmers' problems in collaboration with farmers and, when possible, also with researchers. Nowadays these location specific solutions also become necessary in many irrigated areas where the crop yields are no longer increasing, and where environmental problems endanger sustainability of the farming system. In addition, often new but uncertain market opportunities have opened up. An extension agent should not tell a farmer to switch to vegetable production as a way of increasing income if prices are likely to drop suddenly.

Market changes are one of the reasons why interest in more participatory approaches to extension has increased greatly in the last decade. Another reason is increased attention to problems of rainfed areas where most of the farm population often lives under poor conditions. The sustainability of many rainfed farming systems is in danger because of soil erosion, decreasing soil fertility and increasing population pressure. If we cannot solve these problems very serious social situations may arise, as we have seen in Rwanda in 1994. But it will not be possible to solve these problems without making full use of the intelligence of the local population and without enlisting their cooperation.

A participatory extension approach not only changes the relationship between the Village Extension Agents and their farmers, but also requires a complete change in the culture of the whole extension organization, and often in other parts of the government bureaucracy as well. It cannot be achieved with an authoritarian style of leadership in the extension organization as the Village Extension Agents would be torn between the desires of the villagers and the orders of their losses. This is why a participatory approach is more often used in NGOs than in governmental extension organizations. (See also the discussion of management styles in Section 10.3.)

It is clear that a participatory approach also depends on the culture and social structure of the society. It is more difficult to achieve in an authoritarian culture and in a society with large differences in interests and power among different groups of farmers. It is also more difficult to achieve in a government extension agency which is a part of a large government bureaucracy than in a small NGO (see Section 9.6).

The reader should appreciate that both authors of this book come from cultures which are not very authoritarian and societies which are rather egalitarian. These conditions may have biased their views about the most desirable extension approach, although many extension people in developing countries, mainly in NGOs, share their view.

The following questions should be answered in planning a participatory approach in an extension organization.

(1) What is the contribution farmers can and should make in planning, implementing and evaluating the extension programme?
(2) Which farmers should make this contribution and how should they communicate with the other farmers over their contribution? Do some farmers try to influence the extension programme in a direction which is in their own interest, but not in the interests of less powerful farmers?
(3) Is there a climate which stimulates farmers to come forward with their ideas about the extension programme, and perhaps to disagree with extension agents, researchers and policy makers when this is in the best interests of successful agricultural development?
(4) Do extension administrators know which information their field staff has learned from farmers that can be useful for planning, implementing and evaluating extension programmes, and do they use this information?

9.6 Non-governmental organizations and extension

Non-governmental organizations (NGOs) play an increasingly important role in rural development, and extension education is one of their policy instruments. They often use a participatory approach in their extension programmes and offer farmers an opportunity to participate in the planning of extension programmes of other organizations. However, discussion of their role in the literature is rather confusing because there are very different kinds of NGOs. Major types are:

- membership organizations through which farmers or rural people try to reach some of their goals collectively;
- grassroots support organizations which help rural people in their process of development;
- development support organizations which assist membership and grassroots support organizations through financial assistance and/or technical expertise.

9.6.1 Types and roles of NGOs

Membership organizations

Farmers' organizations are very important for agricultural development in industrial countries but tend to be quite weak in many less industrialized countries. Frequently they are organized by government officers or politicians rather than by farmers. Sometimes these organizations work for farmers, but often they implement government policies or mainly serve the interests of the organizers.

If we wish to stimulate development of farmers' organizations we should decide which of the following roles they will play:

Educational roles

- Organize meetings and courses where extension agents, teachers and researchers discuss research findings and experiences with farmers.
- Organize study clubs where farmers exchange experiences and conduct experiments, often with the help of extension agents and researchers.
- Establish and manage vocational agricultural schools and training centres for farmers.
- Employ extension agents.
- Organize farm youth clubs where boys and girls learn professional and managerial skills mainly through experiences which contribute to their cultural development.
- Publish a farm journal and other publications which provide farmers with information they need for managerial decisions. Educational radio and TV programmes are also possible in some countries.

Commercial and organizational roles

- Organize credit and input supply, as well as marketing and processing of farm products through cooperatives.
- Provide services to members such as artificial insemination, soil testing, accounting and legal advice.
- Organize quality control of seeds and of animals, e.g. a herd book.
- Test the quality of inputs sold by commercial companies and cooperatives.

Management of common property

- Manage communal grazing land, irrigation and drainage projects, roads, etc.

Defending collective interests of members

- Influence government policies, for example price, tax, zoning and environmental policies.
- Influence government agencies in such a way that they provide the services that farmers need, for example by participating in the planning of research and extension programmes.
- Organize public relations for agriculture.

Other roles

- Religious, cultural and recreational roles such as managing temple, music, theatre and sports groups.

It is possible to have separate organizations for each of these different roles, although one farmers' organization often provides several services, such as a cooperative which also teaches its members to produce good quality products. The best combination of roles a farmers' organization can provide depends on the local situation. It is more difficult to manage an organization which performs many different roles than one which performs only one role. Therefore the availability of capable leaders is one factor which determines the desirable combination of roles.

It is also wise to start with only a few or even one of these roles, adding others after the organization has gained strength and experience. It is better to do a few things well than to do many things poorly.

An organization can only be effective if the members have strong common interests. Therefore it is not always desirable for all farmers to belong to the same organization. For example, it may be preferable to have separate organizations for livestock keepers and vegetable growers, for men and women or for farm youth. The better educated and larger farmers tend to be members of these organizations.

The first farmers' organizations in Europe were established over 150 years ago. They have grown in size and influence, but in this process of growth the managers employed by the organization have often increased their influence on policy decisions at the expense of the influence of the members themselves. Several of the cooperative farmers' banks such as Credit Agricole in France, now are among the largest banks in the world. However, few farmers know how to manage a world class bank.

Extension can play different roles with regard to these farmers' organizations. They can:

(1) Teach farmers how to reach their goals more effectively by establishing and managing an effective farmers' organization.
(2) Use this organization as an intermediary for communication with farmers by:
    (a) participating in organizational meetings;
    (b) teaching at courses organized by these organizations for their members;
    (c) writing articles in their journal;
    (d) involving organization representatives in planning extension programmes;
    (e) stimulating the exchange of experiences and information among members.
(3) Work as an employee in the organization's extension service.

It is possible to finance these farmers' organizations through members' fees, profits of the commercial activities of these organizations when cooperatives, or subsidies from the government, development support organizations or foreign donors. Governments and donors may be convinced that they can help farmers more effectively through these farmers' organizations than through their own employees, for example the extension service of this organization may be more capable of gaining the farmers' confidence than a government extension service. Acceptance of this support may make it more difficult for a farmers' organization to oppose government policies.

Most countries have had indigenous organizations for many years. New forms of organizations based on European and American experience have been introduced in this century. Many, but certainly not all, of these recent kinds of organizations have not been very successful, often because they have been organized for the rural people by outsiders rather than in conjunction with or by rural people themselves. This made it impossible to develop organizations which fit in with the local culture and make effective use of the experience of the indigenous organizations (5). The indigenous organizations may have played a useful role in the past, but in a rapidly changing society they also have to change to include those who perform appropriate roles to meet people's needs.

Organizations with grassroots support

Organizations with grassroots support try to reduce poverty among certain groups of the population. An old Chinese proverb which illustrates the development of many of these organizations says that by giving people a fish they can eat for one day, but by teaching them how to fish they can eat for the rest of their life. These organizations started by giving direct aid to people who were in trouble following a natural disaster or a war. They provided 'fishing gear' and taught people how to use it to give more effective long-term assistance. Unfortunately several of these grassroots support organizations have been more successful in teaching people that it is nice to receive free inputs, tools and credit at favourable rates rather than in emphasizing the importance of improving their managerial capacity. Many government agencies have had similar problems.

Organizations with grassroots support can have many different origins, including:

(1) Churches, which feel obliged to give not only spiritual, but also material help and to support human resource development.
(2) Universities, which feel that they should not only develop theories, but also use these theories for the benefit of poor people.
(3) Business firms and their foundations, which feel a moral obligation to use some of their profits for the benefit of poor people and believe they should not make much profit from the poor.

(4) People with an ideological motivation to help the poor, often by opposing government policies.
(5) People with an entrepreneurial spirit who feel that they cannot work successfully for poor people in a government bureaucracy, or that they can and should continue to work for these people after their retirement. They feel that they can work more effectively and sometimes also earn more by starting a grassroots organization, which may not be very different from a consulting firm.

Most of these grassroots support organizations originate in less industrialized countries, but there are also organizations, such as the Foster Parents Plan, in industrial countries which try to provide help in less industrialized countries in this way.

Voluntary contributions provide one source of finance for these grassroots support organizations. To raise these contributions the organizations sometimes try to create the image that their target group is in a very difficult situation which will not improve without outside help. Governments and donors often provide support because they consider grassroots support organizations to be more successful than government agencies in helping the poor.

Many of these grassroots support organizations try to not only give material help, but also to empower their target groups in order to reduce exploitation and to increase their skills in stimulating government agencies to provide the help they need. Extension education often is one of the policy instruments these organizations use in a process of integrated rural development and may include the development of technologies which are suitable for the specific location where the NGO works. Many of these NGOs work with poor people in a complex, diverse and risk-prone environment where the agro-ecological and socio-economic situation is very different from the situation at the research institute. Hence, it is necessary to develop suitable technologies locally. This implies that there are often no separate institutions for research and extension, and that both roles are performed by the same people. Sometimes, but not always, they cooperate closely with the national agricultural research system to obtain ideas on the technologies they might test locally. Their experience can also be useful for giving new ideas to the researchers.

Development support organizations

Development support organizations help membership organizations and grassroots support organizations in developing countries through financial and material support, training and expertise. Most of them are based in industrialized countries where they may be financed partly by members and voluntary contributions, and partly by the Ministry of Development Cooperation. They also may be based in less industrialized countries, mainly to provide training and expertise, but they can also be a channel to transfer donor assistance. Membership organizations in industrialized countries which like to support similar oganizations in developing countries also belong to this category.

9.6.2 Advantages and disadvantages of NGOs

NGOs are growing in importance because it is often felt that they have more advantages than disadvantages when compared with government agencies which perform similar tasks. Donors and national governments therefore direct an increasing part of their funds through NGOs, which increases the possibility that these NGOs will become so dependent on external funding that they try to please their funding agencies more than their target group, thus becoming less effective. There is a large variation among NGOs so that many of the advantages and disadvantages we will mention only hold true for some.

NGOs can have the following advantages:

- they have a less bureaucratic and more participatory style of leadership;
- partly as a result of the previous point, they have a more motivated staff which is able to approach the target group in a more participatory way;
- they are successful at reaching the poor and landless;
- they try to develop location-specific solutions to farmers' problems;
- they result in fewer conflicting roles between educating people and implementing government policies which are not in the interest of these people;
- they give more attention to sustainable agriculture and to ecology;
- they are more competent at facilitating farmers to learn from their own experience and from each other;
- they are more likely to discover new development methodologies.

NGOs can have the following disadvantages:

- staff members often lack competence in production technology;
- their solutions to farmers' problems are sometimes based more on ideology than on research findings;
- they are often small organizations which cannot implement a nationwide programme and are not very successful in disseminating the new methodologies they have developed to other parts of the country. This latter disadvantage can be overcome to some extent by organizing a federation of NGOs;
- the system for donors' decisions on financing NGOs may force them to give more attention to short-term rather than to long-term results;
- they may be expensive programmes because a large number of expatriates are employed and several NGOs work in the same area;

- they can be used by politicians to increase their power;
- leaders may use the available resources for private gain.

We conclude that extension managers should analyse their AKIS carefully in order to discover the strengths and weaknesses of government agencies and of NGOs in extension. This makes it possible to develop an optimal system of coordination and cooperation. For example, it may be in the best interests of the farmers for NGO staff members to participate in training given by Subject Matter Specialists (SMSs) of the government extension service. Such cooperation may require decentralization of decision-making in government agencies because the NGOs usually work in a limited area.

9.7 Chapter summary

Farmers' participation in the planning, implementation and evaluation of extension programmes is desirable, because they have information which can improve these programmes, because it increases their motivation to cooperate and because it improves opportunities for collective decision-making. It also increases farmers' power to influence their own destinies. Clear decisions should be the goal of farmers and their representatives. Similarly extension agents are expected to contribute positively to this decision-making process, including ways in which the farmers' representatives are chosen. Participation cannot be achieved in an extension organization with an authoritarian style of leadership.

NGOs play an increasingly important role in rural development because it is often felt they have more advantages than disadvantages when compared with government agencies which perform similar tasks. Extension education is one of their policy instruments.

☞ Discussion questions

1 How do which farmers in your country participate in the planning, implementation and evaluation of extension programmes?
2 How should they participate?
3 Which changes in the extension organization are necessary to make a successful participatory approach possible?
4 What can government extension organizations learn from the experience of NGOs with participatory extension approaches?
5 How much influence do and should farmers have on the direction of agricultural development?
6 Which role should extension organizations in your country play in the development of farmers' organizations? How can they gain the competence to perform this role properly?

Guide to further reading

Anonymous (1993) People in community organizations, Chapter 5. In: *Human Development Report 1993*. Oxford University Press, New York.

Burkey, S. (1993) *People First; A Guide to Self-reliant Participatory Rural Development*. Zed Books, London. (An excellent analysis of possibilities and problems based on a wide field experience.)

Carroll, T.F. (1992) *Intermediary NGOs; The Supporting Link in Grassroots Development*. Kumarian Press, West Harford.

Cernea, M. (1991) *Putting People First; Sociological Variables in Rural Development*. Oxford University Press, New York.

Chambers, R. (1993) *Challenging the Professions; Frontiers for Rural Development*. Intermediate Technology Publications, London.

Gubbels, P. (1995) The role of peasant farmer organization in transforming agricultural research and extension practice in West Africa. In: *Agricultural Extension in Africa: Proceedings of an International Workshop, Yaounde, Cameroon*. Centre Technique de Cooperation Agricole et Rurale ACP-CEE, Wageningen, pp. 95–120.

Howard, T., Baker, H.R. and Forest, L.B. (1994) Constructive public involvement, Chapter 11. In: *Extension Handbook, Processes and Practices*, (ed. D.J. Blackburn). Thompson Educational Publishing, Toronto.

Pradervand, P. (1989) *Listening to Africa; Developing Africa From the Grassroots*. Praeger, New York.

Srinivasan, L. (1990) *Tools for Community Participation; A Manual for Training Trainers in Participatory Techniques*. Promotion of the Role of Women in Water Supply and Sanitation (PROWWESS), United Nations Development Programme (UNDP), New York.

The Overseas Development Institute has published a large number of studies on the role of NGOs in rural development. A brief summary of this research is given by:

Farrington, J. and Bebbington, A.J. (1994) From research to innovation: getting the most from interaction with NGOs. In: *Beyond Farmer First; Rural People's Knowledge, Agricultural Research and Extension Practice* (eds I. Scoones and J. Thompson), pp. 203–12. Intermediate Technology Publications, London.

10 Organization and Management of Extension Organizations

'Rule 1: The boss is always right.
Rule 2: If the boss is not right see Rule 1.'

10.1 Introduction

The optimal organization of an extension service and its management depends to a large extent on the tasks it has to perform and the environment in which it operates. As this environment is changing rapidly the tasks of extension organizations also have to change. We mentioned several of these changes already in preceding chapters (1). Major changes include:

(1) Demand for agricultural products is increasing rapidly in many countries because of a growing population and increasing incomes. As a result of this growth in income the demand for animal and horticultural products is increasing more rapidly than the human consumption of cereals. In the past much of the growth in production was achieved by cultivating more land and irrigating a larger proportion of it. These options are are no longer available in many countries because of urbanization, soil erosion, pollution and depletion of aquifers. Most future growth will have to come from increased yields and higher efficiency of resource use. These require more competent farmers.

(2) Economic liberalization opens new opportunities for farmers to sell their products on the world market, but it also increases their exposure to international competition. These developments favour the more efficient farmers who are supported by a well organized input supply, marketing, research, education and extension system.

(3) Many present farming practices are not sustainable. Development of more sustainable farming practices often requires collective decision-

making, whereas extension in the past mainly supported individual decision-making.

(4) It has become at least as important for extension agents to help their farmers to decide on new farming systems as to decide on new production technologies. The farmer often needs help to choose between the different options open to him rather than follow an extension recommendation. In other words, transfer of technology becomes less important than increasing the ability of the farmers to make their own choices.

(5) Farmers obtain new information not only from the government agricultural extension service, but also from a rapidly growing range of information sources. Developments in information and communication technologies have opened up many new opportunities to obtain information. Farmers will only turn to their extension agents for information in those fields where they provide more relevant, more reliable and more timely information at a lower cost than other information sources. Research and extension organizations are required in most less industrialized countries which respond more quickly to farmers' need for information and education than they did in the past. Thus extension organizations should themselves use all available sources of information, including their farmers' indigenous knowledge and experience.

(6) There are strong forces towards a change in the financing of extension organizations through privatization and financial support of governments to NGOs.

As a result of these changes most extension organizations ten years from now will have to be organized in quite a different way. Only those organizations which change drastically in structure and in culture will be able to survive in the new and more competitive climate. Major changes are needed in the ways in which extension agents perform their tasks and relate to their farmers. The extension managers will have an important ask in guiding this change process. We cannot indicate clearly which changes are needed in a given situation, but will discuss some principles in this chapter which may increase the competence of extension managers as guides to the change process.

We first discuss conditions which an extension service must fulfil before it can work well. Collaboration between agents is very important and requires extension service leaders to understand personal relations as a precondition for leadership. However, leadership is wider, because the goals of the service must be understood, and means found and made available for achieving these goals, including personnel management techniques and staff development.

Specialists are required in extension organizations to ensure there is good communication between research workers and the general extension agents who have direct contact with farmers. Furthermore, as women play an important role in agriculture, female extension agents often are required to help them effectively.

We will discuss the advantages and disadvantages of combining extension work with other responsibilities, of different organizational settings for an extension service and of privatization of extension services. The T & V System is analysed as a management system developed for conditions in less industrialized countries.

Finally, we point to the urgent need for more research in this area.

10.2 Conditions for the organization of an extension service

The optimal structure of any organization, including an extension service, will depend to a large extent on the function of that organization. Thus the optimal structure for an extension service is different from that for a factory manufacturing a product on the mass production line.

10.2.1 Good communication

An extension organization helps farmers through communication, which means the service requires efficient internal communication itself. The extension service helps farmers to form opinions and make decisions. Hence members of the organization must understand opinion-forming and decision-making processes, especially the problems farmers have with these processes. Information needed to develop this understanding comes into the organization through its members, especially agricultural extension agents who have frequent personal contact with farmers. This information must reach organizational decision-makers, mainly the senior management staff and the Subject Matter Specialists (SMS) (see Section 10.5). The chief task of these people is to help the extension agents work effectively, and they cannot do this without adequate information from the field.

10.2.2 Information

Much of the information needed to solve farmers' problems with their decision-making will come from research, although some will come from other farmers and some from policy makers, for example, that relating to subsidies, price forecasts, etc. Such information should reach extension agents rapidly, accurately and in a way they can use in their contacts with farmers. The SMSs often translate this information in extension messages, making it easier to understand and apply at the local level. There should be a specialist for each discipline who is an important source of information for the extension organization. However, it may take time to develop effective communication channels when new disciplines become important for agricultural development. This happened recently with knowledge about computer use and information technology.

10.2.3 Adaptation to changing environments

The extension organization operates in an environment which provides it with a budget and with manpower, and which in turn influences its goals and those of the farmers it serves. This environment is changing continuously, and the rate of this change is increasing. Hence, the extension organization must be aware of these changes in time to adjust itself or, wherever possible, to influence these changes. The organization will not be able to influence some external factors, such as energy prices for example, but it should try to influence the pre-service training programmes of its staff.

Management has the prime responsibility to note these changes in the environment, but it may delegate some of its responsibility to other staff members. Such information should be communicated to those members of the organization who need it for their decision-making. These may be different persons, for example, for changes in cultural interests of youth than for the development of a good market for vegetables.

The political environment is important for the extension organization, especially in less industrialized countries. Politicians are expected to win rewards on behalf of the constituents who elected them, and especially for those who played a direct role in helping to get them elected. Therefore, these politicians may try to influence the extension organization to deliver some of the rewards. Education in modern agricultural technology is not always recognized as a reward, whereas access to subsidized inputs and cheap credit is. Political influence therefore may act as a force to divert the extension service from its educational role. Hence, it may be very important to show politicians how important the educational role of extension is for the welfare of their constituents.

Politicians from different parties typically argue with each other. Members of one party may try to align the extension organization with their party, partly because, if effective, such an organization will have considerable potential for influencing voters. However, it will lose the confidence of other politicians and their followers if it does align itself with any one political faction. We believe an extension organization should keep out of political struggles, even if this results in less political support for extension work. Politicians would find it dangerous to vote for decreasing the budget of an extension organization which had gained the confidence of most farm families by working hard in their interests.

10.2.4 Motivation of staff

We have stressed the importance of good communication within an extension organization to help it achieve its goals. Motivation of its staff also is very important for the same objective. Supervision of extension staff is much more difficult than in many other organizations because they work in many locations scattered over a large area, and often at many different tasks. The organization will operate effectively only if staff are convinced personally it is important for them to perform their duties well. Furthermore, if they believe in what they do they are more likely to convince farmers to change than if they work because they are told to and because they need the salary. Farmers usually are sufficiently sensitive to understand why an extension officer tells them something.

All extension agents, to be motivated fully, must know clearly what their task is, and must consider it to be important for their farmers and realistic for themselves. They are likely to lose this motivation if they are unable to complete a task in the time allotted or with the transport available. They are more likely to consider the task to be important if information they have provided is used to make clear decisions regarding the extension programme. On the other hand, they will have little regard for a project in which decisions are made on the basis of what they believe to be incorrect information. It is logical for all extension agents to have their own opinions about the direction agricultural development should follow in their area. They will be strongly motivated to assist this development if the extension programme works in the same direction. Hence it is desirable for the extension organization to discuss this direction openly with staff members at all levels.

Motivation of extension agents also is influenced by the reward system of the extension organization, a point we discussed in Section 5.3 on learning psychology and which we will discuss further in the next sections of this chapter.

10.2.5 Flexibility

Communication and motivation are important factors for maintaining flexibility in an extension organization. The organization will have to reach different goals and new target groups in new ways during the future if it wishes to remain ahead of new developments in agriculture. This requires a continuous programme of organization development, as well as a policy which creates opportunities for the personal development of all extension agents, and rewards those who use these opportunities effectively. The extension organization should try to develop a culture of learning in which all staff members contribute to a process of discovering how the organization can adjust itself to a changing environment with new opportunities. The limited budgets available for extension in most countries mean that less important activities must be replaced by more useful ones.

10.2.6 Identification of management problems

We have noticed one or more of the following management problems in agricultural extension organizations in several, but certainly not all, countries:

(1) Staff pay is low compared with other government organizations, whereas extension agents often have to live in difficult conditions far from schools and other facilities.
(2) Career development pathways are totally lacking or severely limited for field staff.
(3) Extension agents are transferred frequently, often for political reasons.
(4) Transport is a major problem, with inadequate budgets for travel allowances and few vehicles.
(5) For field staff there is a lack of support from or liaison with:
    (a) supervisors, who do not always help remove constraints to effective work;
    (b) research workers, who do not use feedback from extension agents when choosing research problems, and who do little research in farmers' fields;
    (c) input suppliers, who do not always ensure an effective and timely distribution;
    (d) mass media;
    (e) trainers, who do not always stress the practical aspects of their subjects; and
    (f) other government departments involved in rural development.
(6) The walls between different government departments can be high, which makes cooperation between them difficult. Many rural people receive their income from different branches of agriculture as well as from sources outside of agriculture. The present structure of government makes it difficult to develop an extension programme which helps them in an integrated way.
(7) Extension agents have to spend excessive time writing reports and collecting statistics which are seldom used for improving rural development.
(8) The extension programme is not adapted to varying local conditions, or there is no suitable technology for solving farmers' problems.
(9) Part-time farmers and women have limited access to extension agents.
(10) Extension agents are not always given realistic tasks, and are not held responsible for accomplishing these tasks. Often there are no clear job descriptions.
(11) In a large number of countries, the Ministry of Agriculture is obliged to find employment for all graduates in agriculture. Therefore, nearly the whole extension budget is used for salaries, leaving very little for operating expenses, thus preventing the staff from working effectively.

Furthermore, it is impossible to select the most suitable extension agents.

(12) Technical assistance projects sometimes have developed their own extension organization in such a way that it cannot be integrated with the national extension organization. This is often because far more is spent per farmer on extension work than is available for the nation as a whole.

10.3 Leadership in extension organizations

An extension organization will try to attract staff who are willing and able to contribute as much as possible to fulfilling organizational goals. Individual staff members will accept positions within the organization because they expect to achieve a large proportion of their own goals in this way. It is seldom that both sets of goals are in complete agreement. Even if they were when the staff member joined the organization, they may diverge afterwards as the goals of the individual and the organization change.

The leadership of an extension organization will try to ensure its staff members contribute to achieving organizational goals as much as possible. They can do this by indicating clearly what each extension agent should and should not do, as is done in the T & V System (see Section 10.9). They also can give the extension agents extensive freedom in deciding what should be done to help farmers. Such an approach is most effective if these decisions are discussed at meetings in which staff try to clarify with management what the best decisions are in different situations. This leadership style, termed as participatory style, is favoured by the extension organizations in many industrialized countries. Clearly, there are many alternatives which lie between these two extremes.

The alternative which is to be preferred depends on a number of factors:

- environment
- expertise
- applying the correct leadership style in the given situation
- leadership power
- time available to superiors for their leadership task
- common interests

10.3.1 Environment

Management can prescribe what each extension agent should do if the situation in which the extension service operates is very uniform and predictable. The same is done for workers on an assembly line where much of the work could be done by a robot. An extension agent's situation usually is much less predictable and becomes increasingly so, as we discussed in Sections 9.5 and 10.1.

Leadership style will influence the relationship between extension agents and their farmer clients. They will be inclined to give directive advice if their superior officers act in an authoritarian way towards them. On the other hand, if their superiors discuss the best way of doing their work they will approach farmers in the same non-directive way.

10.3.2 Expertise

The extent to which extension agents or extension managers have the expertise needed to solve farmers' problems is important. A surgeon is likely to have more expertise than his hospital administrator in how to operate on a patient. The same increasingly is true in extension organizations as extension agents become more professional and specialized, especially in industrialized countries. These organizations sometimes prefer managers who are not experts in local agriculture so that they can concentrate on management rather than on extension. In some less industrialized countries extension agents with a low level of technical expertise sometimes have been appointed who might make serious mistakes if not told exactly what to do. However, the expectation that they are incapable of making good decisions about their work may be a self-fulfilling prophecy. If the leaders believe they must do the thinking for the extension agents, they may stop thinking.

Clear decisions must be made regarding content and timing when different methods supplement each other in an extension programme. The probability these decisions will be carried out properly increases when the whole staff of the extension units participate in the decision-making. In the case of individual extension it is possible for each extension agent to decide on his own message. Management must be sure that these individual messages do not conflict with each other as that would confuse farmers.

10.3.3 Applying the correct leadership style in the given situation

Extension agents' expectations regarding leadership style can vary. There has been a strong tendency in the last 30 years towards developing a more democratic style of leadership in most industrialized countries. This is related to changes in child rearing and educational practices which try to develop children as people who can think for themselves and find their own way. Children are taught much less than previously about what their parents or teachers think is good for them and for the society.

The structure in many, but not all, less industrialized societies is much more hierarchical. The boss in these societies is expected to decide what his or her staff should do. A manager who uses a Western style of democratic leadership might be seen as weak and incapable. The subordinates often will work hard to enable their superior to do a good job, but they may expect this superior to help them achieve promotion (2).

10.3.4 Leadership power

The leaders' power over their staff is often limited in extension organizations because the extension agents have many opportunities to prevent their superiors from obtaining an accurate picture of how they work. Furthermore, if these are civil service positions, it may be difficult to dismiss somebody who does a poor job, and rather difficult to promote those who do a good job.

10.3.5 Time available to superiors for their leadership task

The authoritarian leadership style, with its limited delegation of decision-making, has the disadvantage that leaders become overburdened with decisions and hence delay making them, or make them without adequate consideration of all aspects. This appears to be an important cause of a rather inefficient bureaucracy in some countries. It is often the rule in industrialized countries for decisions to be taken at as low a level in the organization as possible to free top management to concentrate on making the most vital decisions for the organization.

The participatory style of leadership requires more time for decision-making than the authoritarian style. It therefore is unsuitable where decisions have to be taken very rapidly as, for example, in a locust attack. On the other hand, implementation of decisions is much faster with a participatory style because during the decision-making process most extension agents already will have understood what they are expected to do, and voluntarily will try to do it.

10.3.6 Common interests

The extent to which the extension agents and their organization share common interests is variable. Much academic writing about organizational leadership assumes implicitly that there is always a shared interest, but we believe it is an unrealistic assumption. A change in the environment may force the extension organization to take on new responsibilities for which a number of extension agents cannot be retrained. For example, the number of poultry farms in the Netherlands has decreased drastically, whereas the number of flower growers has increased. Therefore the extension service has decreased the number of poultry extension agents in order to increase the number of extension agents dealing with flower production. The poultry extension agents cannot be expected to share the responsibility for this decision. Such a decision should be made by management which should listen seriously to the arguments of all its staff members, while being prepared to make a decision which is against the best interests of some staff members.

Each extension manager will have to choose his or her own style of leadership, taking into account these six points. The choice will vary on the participatory–authoritarian continuum according to the situation, but usually it will be somewhat more on the participatory side than in most other government organizations.

Research in Kenya has shown that an authoritarian style of leadership in a unit of an extension organization correlates with low productivity of extension agents (3). However, it is not certain this will hold true everywhere. Research in factories in industrialized countries has shown this style has an adverse effect on worker satisfaction and on internal communication in the organization, but no clear effect on productivity.

We have shown that the success of an agricultural extension organization depends firstly on the relationship between the extension agents and the farmers. All other actions in the organization are important mainly because they strengthen this relationship. Therefore it is important to reward extension agents' good work both by showing appreciation and by increasing their salaries. In many industrialized countries the salary levels for extension agents and research workers with comparable levels of education and years of experience are about the same. We consider this is desirable for less industrialized countries also, otherwise it will be impossible to attract the capable and motivated extension agents needed by the country. In any case, the organization may decide to pay a hardship allowance for working in remote areas far from schools and other facilities.

It is often assumed that the more managers of an extension organization are influenced by their staff, the less influence they will have on their staff. This assumption is not correct. Their staff will have more confidence in the decisions of the managers when they listen to their views than when they take arbitrary decisions which, in the staff's opinion, are based on incorrect information. The confidence of their staff determines to a large extent how much influence the leaders have. There are limited possibilities for leaders or staff to influence each other in an organization where people distrust each other. This is perhaps even more the case in an extension organization than in many others, because it is difficult to get an accurate picture of the role behaviour of all workers.

Our analysis of the management process must involve a study of which decisions are made, as well as the processes of making and implementing these decisions. For example, we should analyse whether the goals are clearly formulated and based on a systematic analysis of the situation and of the available resources. Management has limited influence on the implementation of these decisions in many extension organizations. Field extension agents put them into practice, but seldom are observed or supervised by managers who are busy with administrative paper work and office meetings and may lack transport to go to the field.

In an organization it is necessary to make strategic decisions on changes in the role of the organization, decisions on the management of resources, e.g. on budgets, and on the way the work is actually performed. This requires short- and long-term decisions and the implementation of these decisions. These should be well attuned to each other. We have the impression that many long-term decisions receive insufficient attention because they can be delayed without causing immediate problems. This is especially true in organizations where staff are frequently transferred. In these situations the successors to the present managers will have to solve the problems. Otherwise they might be rewarded for the good planning of their predecessors.

Agricultural development depends not only on farmers and the extension organization, but also on many other organizations, such as departments responsible for irrigation, research, credit, input supply, public health, farmers' unions, etc. Hence we must also pay attention to management of the joint efforts of all these organizations as well as to extension. A major problem for agricultural development almost without exception is that many of these organizations are not really rewarded for cooperation in promoting the farmers' best interests.

10.4 Staff development

Its staff members are the most important resource available to an extension service. Different and usually higher standards are expected of them under conditions of change. Therefore, managers must ensure the quality of this resource is improved as much as possible. They can contribute to this improved quality with a leadership style aimed at staff development through the manner in which they direct the staff, by motivating their staff to work for development of their farmers, and with a programme of systematically organized training courses.

Staff development is a crucial element in the process of changing an extension organization. Suppose that at present the extension agents act more or less as postmen, who pass the messages they receive during their regular training from the SMSs to the farmers without changing anything in these messages. Later we become convinced that in a rapidly changing extension environment they should function in a more participatory manner in which the extension agents become the farmers' partners who discuss the various options open to them to increase their income in a sustainable way in their specific situation. This requires extension agents who are much more competent to diagnose the situation of their farmers together with them, and who are capable of taking new initiatives to develop new solutions to these problems. The leadership style of their superiors and the way in which the SMSs support extension agents also has to change drastically.

Antholt (4) is convinced that the following skills are required by an agricultural extension agents in the 21st century. Ability to:

(1) work under complex and fluid circumstances with little supervision;
(2) diagnose farmers' problems, and the willingness to do so effectively;
(3) listen to and learn from farmers, and the willingness to do so;
(4) communicate effectively with farmers and farmers' groups; and
(5) present options, based on principles of science and good agricultural practices, which widen the real choices available to farm families.

It cannot be denied that most extension agents do not currently possess these skills. It is arguable whether Antholt gives enough attention to the variation in situations in which extension agents will work in the near future. Clearly in Eastern Zaire and Central Thailand, for example, different skills will be required.

This changing role of extension requires not only a massive retraining programme of the whole extension staff, but also a different training style which is no longer limited to passing on messages, but which tries to develop creativity. For example, most of the lectures may be replaced by opportunities for trainees themselves to discover new information in a library or through experimentation and work with case studies in which the trainees are provided with opportunities to develop new solutions for actual field problems.

Such changes in the contents and methods of staff development cannot be achieved overnight, partly because of lack of trainers who are able to give this kind of training. Staff members of donor agencies do not always appreciate that it may take many years before these changes in staff development can be fully implemented.

Leadership style has been discussed in Section 10.3. The participatory style of leadership is directed towards promoting staff development and motivation. An extension manager can try to help the extension agents find solutions to their problems in a non-directive way, just as an extension agent can help a farmer. This type of help can stimulate their personal development so they will be in a position to find their own solutions to problems in future.

Managers must know well what their extension agents have to learn if they are to use this style successfully. They must make regular field visits to observe extension agents and to talk to farmers in order to learn which skills their extension agents must develop. Showing them how they should do a certain job also can be effective. However, this must be done carefully so that farmers do not lose confidence in their extension agents. It is better to discuss possibilities for improvement in private rather than in front of farmers.

10.4.1 Rewards

As we discussed in Section 5.3, it is important for their development to reward extension agents for doing a good job. If they see that colleagues are promoted who do their job in a certain way, they will be inclined to do it in the same way. Thus they might learn that they have to work hard and long hours to solve their farmers' problems, or that they have to pay considerable attention to their boss when he visits the field.

It is not only material rewards which are important. People may feel rewarded if they are convinced they have made a useful contribution to an important goal or have completed a task successfully which others had failed to complete. Most extension agents need to discuss regularly how they have done their work, what they have done well and what they might have done better, and how they can learn to improve their performance. There is a greater need for extension agents to discuss these points than there is for many other employees, because it is difficult for them to judge the quantity and quality of their work. For example, teachers can compare their own students' examination results with those of other teachers' pupils.

10.4.2 Training

All extension services require a systematic in-service training programme, but it is especially important in those services which have had to attract extension agents with a rather low level of competence because of the lack of well-trained agriculturalists in their country. Two types of training are desirable:

(1) Regular training at staff meetings to ensure agents are capable of performing their work satisfactorily in the next few weeks. This training may be given mainly by the SMSs and extension managers.
(2) A series of short courses to increase agents' competence in specific fields. These courses may be given both by SMSs and by the staff of training centres. The courses may focus on an aspect of production technology or an extension methods.

Extension agents do not only learn from courses, but all good extension agents try to learn in many different ways how they can perform their tasks better. For example, if they work in a vegetable growing area they may decide they should learn more about vegetable growing even if that is not included in their organization's training programme. In such a situation they may turn to their colleagues, to the literature, to successful farmers, researchers or their superiors to find a way to meet this challenge. These self-directed learning projects are often more important for the success of the extension service than the courses and the extension agents are more motivated to learn from a learning project which they have chosen themselves than from a course in which they are obliged to participate. Therefore extension management should encourage and stimulate those self-directed learning projects, for example by helping extension agents to find and gain access to relevant sources of information and to motivate them to learn from each other. The extension agents' demonstrated ability to learn new knowledge and skills which are important for their work could be a criterion for promotion (5).

Courses must be taught in a practical way. Teaching in many schools and universities tends to be rather theoretical, but extension agents must be able to demonstrate as well as talk about a practice they wish to convince a farmer to change. Hence, much of the training should be given in the field rather than in the classroom.

It is often assumed that extension agents do not work properly because they lack training. However, the reason often must be sought elsewhere, for example, in poor planning or supervision, or in a reward system which does not reward good work. People, including extension managers, are inclined to look to others for the cause of a problem rather than to themselves.

10.4.3 Motivation and job satisfaction

Motivation of production workers and professional staff is an important but rather contentious issue in modern management theory. Although writers disagree about the components of management systems which motivate workers effectively, at least one writer has proposed two main types of factors which stimulate or suppress work activity (6). Achievement, recognition, responsibility, advancement and the work itself all determine job satisfaction. On the other hand, organization policy, supervision, salary and working conditions can cause job dissatisfaction.

Factors involved in producing job satisfaction thus are separate from those causing dissatisfaction. The first set of factors lead to growth and motivation, while improvement in the second set merely reduces discontent. Hence, managers should provide employees with opportunities for advancement and recognition, while at the same time minimizing causes of dissatisfaction such as lack of policy, poor supervision, low salaries and bad working conditions.

10.5 Specialists and generalists

Farmers with a range of different problems usually need extension assistance which may be provided by one service or by several specialist services. A single extension agent cannot be expected to master all the expertise required in a service which deals with a very wide range of problems. There are over 30 000 academic research workers working in Indian agriculture. A local extension agent with an agricultural college diploma cannot be expected to know as much as the combined knowledge of all these research workers. It must be possible for a generalist extension agent to operate in an organizational structure in which he can draw on knowledge from these research workers as required. The most practical solution is to have specialists in the extension service who are aware of research findings in their specialty as well as farmers' problems. They then are in a position to say how the research can be used to solve these problems. The catch is that many problems require cooperative inputs from several specialists. Cooperation becomes extremely difficult if each specialist considers his own subject to be more important than the other subjects.

There are two types of Subject Matter Specialists (SMS):

(1) SMSs who are responsible for a certain branch of agriculture, e.g. horticulture.
(2) SMSs for a certain discipline, e.g. plant protection.

The task of a *specialist* within an extension service is:

(1) to keep generalists aware of developments in their special field by lectures, publications, etc., as well as with systematic in-service training;
(2) to support the generalists when solving difficult problems; this can be in the form of on-the-job training for the generalists, or as a service to the generalists in the case of rare and difficult problems, by solving these problems for them;
(3) to keep the research workers aware of the farm problems associated with their discipline for which no satisfactory solution is available yet, and of farmers' reactions to and modifications of solutions developed in the past;
(4) to integrate different research workers' knowledge, the scientific literature and farmers' experience into practical recommendations;
(5) to make extension programme planners aware of farmers' problems which can be solved by the specialist's specialty;
(6) as part of the extension programme, to cooperative in educating farmers via the mass media, demonstrations, talks, preparation of extension aids, etc.;
(7) in the case of a branch SMS, also to analyse trends which are relevant for the development of the branch, for example, those regarding markets.

The task of the *generalist* when cooperating with specialists is:

(1) to integrate different specialists' knowledge into practical recommendations;
(2) to make use of the correct specialist(s) when solving a practical problem;
(3) to take care that the specialist does not exaggerate;
(4) to make the specialists aware of practical problems which require a solution.

An alternative to having specialists within an extension organization is to have specialized extension organizations for different branches of agriculture. There might be an organization for extension work on crops or even several organizations for different crops, for animal husbandry, for social forestry, etc. It is difficult in these conditions to develop an extension programme in which the mass media and interpersonal communication are well integrated (see Chapters 5 and 6) and which takes the needs of the farmers as its starting point (Chapter 7). The farmers might get confused if they are approached by so many different extension agents who often given conflicting advice. Also, an individual farmer might find it difficult to identify those extension officers who are supposed to help. As a rule we believe it is preferable for different agencies to delegate their extension task to one extension organization which can plan overall development of the farm.

An exception might be made in the case of highly specialized farms. Here it may be desirable to work with field level extension agents who are specialized for the major types of farms. Otherwise it is difficult for the extension agents to know more about this type of farming than the farmer knows himself. This is why there are specialized extension agents in the Netherlands for bulb growing, cut flowers, egg production, etc. A flower grower will grow only flowers and certainly will not produce eggs.

About 18 per cent of the extension agents in industrialized countries are SMSs, whereas in less industrialized countries only about 6 per cent are specialists. The percentage is low in these latter countries mainly because there are few well trained specialists available. This shortage also may contribute to poor standards of training for the extension agents.

10.5.1 Information specialists

Extension organizations not only have specialists for different technical aspects of agriculture, but also for preparation of information materials and training programmes.

An effective extension programme needs the support of leaflets, radio programmes, audio-visual aids, etc. Preparation of these materials requires an expertise which most extension agents do not have. It also may require rather expensive equipment which can be used more effectively in a specialized agricultural information unit. The staff of this unit can build good relationships with the gatekeepers in the mass media in order to place their materials in newspapers, farm magazines, radio and television programmes. One study of a number of extension organizations showed there was on average one staff member in an agricultural information unit for every 139 extension agents (7). The extension organization probably would be more effective if more information specialists were employed.

Agricultural information units provide materials which support the extension programme. They should not try to develop their own extension programmes. Therefore they must be informed well in advance about which topics the extension programme will focus on, as preparation of extension materials takes time. The content of these materials usually will be decided by the SMSs. The information specialists are responsible for making it more readable and more attractive to the farmers. Normally, they would check their draft manuscripts or scripts with the SMS to minimize the risk of publishing or broadcasting any mistake.

Information specialists should have close contact with the target group to ensure they make materials which their audience needs and understands. They do not have an office job. Pre-testing and evaluating their materials can be one of the ways to keep in contact with their audience.

The training specialists will be responsible for the training programme we discussed in Section 10.5. Most training will be done by technical SMSs, although some might be done by research workers or teachers in agricultural schools. The training specialists usually will not do much of the training themselves, but they will develop an effective curriculum, organize who contributes what and when to this curriculum, discuss the training needs of the trainees with the trainers and ensure that training materials and facilities are available at the right time.

10.6 Female extension agents

Considerable attention has been given in recent years to the fact that a large proportion of agricultural work in many countries is done by women, whereas in most of these countries only a small proportion of the agricultural extension agents are women. This imbalance can make it difficult to reach such an important target group.

Extension organizations wishing to support the role of women should analyse which activities men and women perform in agriculture in their area, what access both groups have to different resources, who benefits from production and controls income, what the information needs of farm women are and through which communication channels this information might be provided (8). There are large variations between cultures in the way different members of the family contribute to agricultural production and to consumption. Negotiations between husband and wife may be needed regarding each contribution to agricultural production and the distribution of the benefits from this production.

10.6.1 Major roles farm women play

(1) Usually there is a division of labour whereby certain tasks are performed by men and others by women. Men are often responsible for land preparation and women for feeding animals in the compound, seed

selection and storage of products. It is also possible that women are responsible for certain crops or animals and men for others. Even if women perform most of the labour, men may have more responsibilities for decision-making, partly because they are usually in charge of the money. However, in some East Asian countries and some parts of Europe this is the women's responsibility. Even if the men are formally responsible for decision-making, many decisions in fact are made by the women. For example, the men may not know how much of which feed the women give to different animals.

(2) Men and women have separate enterprises, for example, the women are responsible for all decisions regarding small animals (chickens, goats) and men for those regarding cash crops. In this situation the women can usually keep the money they earned from their enterprise. Often research and extension give less attention to the women's enterprises than to those of the men.

(3) Women perform certain tasks in their husbands' fields, but they also are responsible for their own fields to grow food crops for their family. This is the pattern in much of Sub-Saharan Africa.

(4) Women have the full responsibility for their farm, although they may get some assistance from male relatives for land preparation. This can be because they are not married, widowed or divorced, but increasingly it is because the men have gone to the cities to look for a more highly paid job, a process which makes extension work to women farmers more important than it was in the past. Often the farms directed by women are smaller than those run by men.

10.6.2 Contact with extension agents

Female farmers usually have less access than male farmers to inputs, credit and other resources. This implies that extension messages which are helpful for male farmers may be useless for female farmers. They may even be harmful, for example, messages about mechanization of land preparation may increase the size of the fields the women are supposed to weed, thus increasing their workload, which usually is very heavy in this period already.

Male farmers as a rule have much more contact with extension agents than female farmers. This may be because of cultural constraints against male extension agents working with female farmers, because male farmers are responsible for the cash crops to which the extension service often gives more attention than to the food crops, or because female farmers are less educated. In several cultures women are not supposed to speak in a meeting unless asked. Such constraints mean that many female farmers have less exposure to the outside world than men. Hence they do not know as well as male farmers what problems they could ask for help for from extension agents.

10.6.3 Appointing female extension agents

It is often proposed to appoint more female extension agents to solve such problems, a solution which works quite well in some cultures. For example, in some parts of the Philippines most of the extension agents are females with a BSc in Agriculture. Many of them are married to a local farmer and are well able to integrate knowledge from research and from farmers' experience. In their culture it is acceptable for men to listen to the advice of a woman, although it is more difficult for a woman to convince farmers that she can give reliable information than it is for a man. In many other cultures women are not expected to give advice to men.

There are other difficulties. It is often impossible in many countries to find enough women with the level of education required for an extension agent. Home economists may be available but they are seldom trained well enough in agriculture.

Many well educated women are not willing to work in rural areas because their husbands have jobs in the city. Others are not yet married, but realize that they have less chance of meeting a well educated man in a village than in a city.

It also may be culturally unacceptable for female extension agents to use the most convenient kind of transport such as bicycle or a moped, or to stay out for a night to meet farmers at a convenient time.

It is also possible to stimulate male extension agents to give more attention to working with farm women. Working with women's groups can often be effective, partly for the reasons we mentioned in Section 6.2.3, but also because it is possible to reach more women in this way. There are fewer cultural constraints for a man to work with a group of women than with individual women whose husbands are elsewhere. It can be expensive employing separate staff for male and female farmers.

10.6.4 Problems to resolve

An issue which causes some arguments is whether extension on agricultural and home economics topics should be given to farm women by the same person or by two different persons. Teaching by one person has the advantage that it becomes easier for this person to gain the women's confidence, to work with women's groups and to prevent presentation of conflicting information in the fields of agriculture and home economics, for example in demands for labour. However, it will not always be possible to find extension agents who are able to teach both topics. Few male extension agents are competent to teach home economics subjects, and it may be difficult for them to gain the women's confidence sufficiently for them to teach nutrition, for example. Female extension agents are often scarce and not always adequately trained in agriculture. About 40 per cent of the time training home economists in Kenya is spent on agriculture and 60 per cent on home economics, because it is appreciated that they could be more helpful to farm women if they can provide help in the home as well as with the farm. Experience has shown that they do a good job with the support of SMSs in agriculture and animal husbandry, an idea which could also be considered in other countries. There are some countries where better educated village girls and widows have been given some additional education to enable them to work as assistant extension agents for farm women.

Working with women may require different extension methods to those used when working with men. There are women's groups in many countries which can play an important role in reaching women. Experience in Zambia has shown that few women are able to leave their home to participate in a residential course at a Farmers' Training Centre, but many participate in a mobile course in their own village. It may be easier to reach the women in their local vernacular than in the national language because of their low level of education.

The extension service must have information which is useful for them if it is to work successfully with farm women. The agricultural research programme may have to be redirected towards paying more attention to women's crops, animal nutrition, post harvest technology, etc.

Great care must be taken in trying to liberate rural women from the suppressed position sometimes identified by female experts from Western countries. Such moves may create much resistance among their menfolk. In the long run it may be more helpful for the women's cause to adopt a slower approach in which people from the country itself influence the desired relationship between men and women.

According to Western cultural values, and even according to the law in several industrialized countries, female extension agents should have the same opportunities for promotion to supervisory positions as their male colleagues. These values are not acceptable in all other countries, although in some less industrialized countries such as India and the Philippines a large proportion of women occupy supervisory positions.

Decisions regarding the role of female extension agents should be based on a sound understanding of the culture of the country, and often on the culture of a specific region or tribe as well.

10.7 Combining extension education with other tasks

In Chapter 2 we pointed out that extension education often is only one of many tasks performed by rural development workers. They may be responsible also for distribution of inputs and supervision of credit. Is this desirable?

Many countries have received a loan from the World Bank to expand and improve their extension services. This loan often is granted on the condition that extension agents are relieved from all tasks except teaching farmers how to increase yields and their income. However, the Bank has become somewhat less rigid in applying such conditions, because it has been recognized that the most desirable structure of an extension organization depends on the local situation.

10.7.1 Advantages

There are several advantages to combining extension education with other tasks.

(1) It makes coordination of different development activities easier. There is no point in teaching farmers to use fertilizers if none are available, or to grow product for which there is no market. In early stages of rural development the extension agent may be the only person available to distribute seeds of improved varieties, fertilizers and pesticides. Furthermore, information about changes to its pesticide recommendations by the extension service is vitally important for pesticide distributors. Such coordination also can be achieved in regular meetings.
(2) Traditional farmers often do not recognize the need for extension advice, whereas they recognize the need for inputs or credit. Therefore it is easier to reach these farmers with a combination of tasks. Furthermore, when visiting a farmer to distribute inputs, the extension agent may identify a disease or other problems which the farmer has not identified. Thus the extension agent is able to provide more timely advice or assistance.
(3) It can reduce travel costs. This is very important where costs are high because of low population density, or travel difficulties due for example to mountains as in Nepal. It is also important where production per farm is low, perhaps because most farmers work part-time.

10.7.2 Disadvantages

However, there are important disadvantages in combining agricultural extension with other tasks performed by the same person or the same organization.

(1) It is almost impossible for the extension agents to gain the farmers' trust and therefore be effective where they also have a policing role such as prevention of disease (see Section 3.1).
(2) If extension agents have to enforce repayment of credit they are unlikely to press farmers very hard in case they lose their trust. The rate of credit repayment tends to be very low under these conditions.
(3) These other tasks take time which cannot be spent on extension work.

Administrators tend to give extension agents all kinds of field tasks because they are in the field anyway, although sometimes the administrators fail to recognize the costs of diverting extension agents from their principal task. Farmers may lose their confidence and trust in the service, or develop a poor image of the extension agents. It is often easier to delay extension activities than other tasks such as collection of statistics or drawing up accounts of inputs distributed. The net result is that little extension work is done when it is combined with other tasks.

(4) There can be conflict of interest between the marketing and credit organization and the farmers. For example, the organization might have a large stock of a certain pesticide which it would like to dispose of, whereas it also sells other pesticides which are more profitable or less dangerous for the farmer.

(5) Government bureaucracies often are not very efficient at distributing inputs or marketing farm products. This inefficiency may create serious problems for the farmers; for example, the fertilizers may not be available on time. If the extension agents are responsible for this distribution, they are likely to lose the farmers' trust, and hence their effectiveness.

(6) These other tasks may require agents with different personalities and an organization with a different organizational structure from that of a conventional extension service. It is preferable to create a special unit for the purpose in the situation where an extension organization also is given regulatory duties. This prevents role conflicts for the extension agents which may result in loss of farmers' trust.

10.8 Organizational setting and privatization of extension services

In our discussion of the AKIS (Section 2.4) we noticed that agricultural extension can be provided by many different sources. In this section we will discuss where it can be organized within the government, how the government can support it elsewhere, and we will mention a number of other organizations which provide extension to farmers. This is followed by an analysis of the present trend towards privatization of governmental extension services.

In most countries the agricultural extension service is one of the departments of the *Ministry of Agriculture*. The Ministry uses the service as one of its instruments to realize its agricultural development policy (Section 2.2). If agricultural extension and research are organized in the same department, linkage between research and extension becomes relatively easy. For example, the SMS can be given an office in the research institute of his specialization, which stimulates formal and informal contacts with the researchers.

A good linkage also can be achieved between separate departments, if everybody accepts that they are jointly responsible for supporting agricultural development. In a few countries agricultural research is organized in the Ministry of Science which can make a good linkage difficult. Also, research done in the agricultural faculties of the Ministry of Education is not always fully utilized by the extension service.

In some countries, such as the USA, agricultural extension is provided by the *agricultural universities* or by vocational agricultural schools. This system has worked well in the USA, but most attempts to introduce this system in less industrialized countries have failed, because the Ministry of Agriculture feels that it cannot fulfil its responsibility for agricultural development without an extension service. In many of these countries the agricultural universities are responsible for extension to farmers in a small area, where they can test new methods. However, these methods seldom are widely used by the governmental extension service, partly because a good linkage between these experimental programmes and the governmental extension service is lacking, and partly because the university can use more qualified manpower and other resources than are available on a national scale.

Vocational schools can play a useful role in organizing short courses for their former students now working in agriculture. What they have learned in school will be out-of-date long before they retire some 50 years later. Sometimes these courses are organized in cooperation with the extension service.

In some irrigation and settlement areas a government agency is responsible for the *total development of the area*, including the engineering works, provision of credit and supplies, marketing of products and extension. This makes it possible to work with settlers with little farm experience, such as unemployed youth. A high level of agricultural production cannot be expected until the settlers have learned to manage their own farms.

In Francophone African countries there are organizations responsible for the *total development of a certain crop*, including provision of inputs and credit, marketing and extension. We find the same type of organization for export crops in some other countries as well. This system may limit farmers' freedom to make their own management decisions. In some countries the system has had a favourable effect on production of the crop involved, such as tea in Kenya. In several countries it has limited the production of other food crops grown by the same farmers because they received little or no professional help for these crops. Furthermore, they are stimulated to use nearly all available resources for the development crop.

Agricultural extension and information is increasingly provided by:

- commercial firms selling inputs to and buying products from farmers;
- banks;
- private consultants working for a fee;

- accountants, who used to keep records for tax purposes, but now use these data also for farm management advice;
- publishers of farm magazines and other publications;
- computerized information services and data banks.

10.8.1 Privatization

In recent years there has been a trend towards privatization of governmental extension services. Farmers are expected to share the responsibility for this service and pay all or part of the costs. Reasons for this change include the following:

(1) Budget deficits make it difficult for the government to pay for such a service.
(2) It is hoped that by making extension agents accountable to farmers who are competent to judge the quality of their work, this in turn will make the extension service more efficient.
(3) In a number of countries many farmers have serious doubts whether the Ministry of Agriculture tries to serve their interests rather than the interests of urban people or of politicians and their relatives and friends. It is difficult in this situation for extension agents employed by the Ministry to gain farmers' confidence, although without this confidence they will have little impact among farmers.
(4) Farmers are the main beneficiaries from extension service activities, and therefore it is fair that they pay the costs.

This last argument may not be correct. A good extension service improves the efficiency of agricultural production, and brings about lower prices for agricultural products through competition, unless the prices are mainly determined by agricultural production in other countries or by government price policies. Therefore the extension service often will have more impact on decreasing the cost of living for consumers than on increasing farmers' incomes from the market place. In many less industrialized countries most of the consumers are farmers, their family members and their labourers, all of whom may benefit from a good extension service.

How farmers can contribute

There are different ways in which farmers can contribute to the costs of a privatized extension service:

(1) They can pay a fee for each visit an extension agent makes to their farm, or for each other service the extension agent provides, which is also the way consulting firms are paid in many other branches of industry.
(2) A levy can be charged on certain agricultural products from which agricultural research and extension are financed. However, it is easier to

collect from some products than from others. In less industrialized countries it is much easier to collect it from export products than from produce sold on local markets, especially when a marketing board has a monopoly.

(3) Costs can be met from membership fees paid to a farmers' association. The problem with this approach is that not all farmers will belong to the association. As we outlined in our discussion of the diffusion of innovations (Section 5.6.4), non-members also benefit from the extension service through informal communication with members who are visited by extension agents.

(4) The extension service can receive a specified portion of the extra income a farmer earns as a result of advice given by the extension agent, for example from a yield increase greater than the regional average. Such a scheme requires a reliable farm accounting system which cannot currently be achieved in many countries.

Advantages

Advice from a privatized system may be more effective because the farmer can select an adviser who is best able to help. The farmer also is likely to prepare questions more carefully in order to make best use of the adviser's time for which the farmer has to pay. Furthermore, the former might be more inclined to follow advice which he or she has paid for in the first place. It is also unlikely that agents in a privatized extension organization have roles in the implementation of government policies which are incompatible with their educational role.

Disadvantages

Privatization of the agricultural extension service has the disadvantage that it may hamper the free flow of information. Government extension agents often contribute to farm magazines and agricultural radio programmes either without asking for payment or for only a small honorarium. With privatization they may be inclined to charge for their services in order to earn the money their organization needs for its survival. Farmers also may be less inclined to tell their colleagues what they have learned from the extension agent as they do not like 'free riders'.

Extension agents in a privatized extension service are inclined to focus on the larger farmers who can afford to pay their fees. Subsistence farmers will seldom be able to pay this fee. Although publicly funded extension services also tend to pay more attention to large rather than to small farmers, it is likely that this tendency is stronger in a privatized service. The private extension service will also concentrate on topics for which farmers are willing to pay a fee. For example, farmers will be more inclined to pay for information on the right pesticide to cure a plant disease than for education about how to prevent degradation of the environment, even though the second topic may be more important for them in the long-term. Extension agents will also select extension methods which make it possible to recover their costs. It is easier to charge for a farm visit than for attendance at a result demonstration, although the latter is often more cost effective (9).

Sharing the cost

In some countries farmers' associations and the government share the cost of the extension service. The associations employ extension agents because they see this as an effective way of increasing their members' incomes. In several countries the government subsidizes extension activity by the farmers' association, whereas in other countries the extension agents are employed through an organization established jointly by farmers' associations and the government. Government in turn may be the Ministry of Agriculture or local government. Extension agents are under more direct farmer control if these costs are paid by local rather than by the national government. However, there may be more political interference in the extension service if local government is involved. For example, extension agents may be expected to favour members of the party in power or be expected to support this party in an election campaign.

It will have become clear that privatization of a government extension service has advantages as well as disadvantages. The local situation will determine whether or not the advantages outweigh the disadvantages. Also it should be clear that no general rules can be given for the best way to implement privatization.

10.9 The Training and Visit System

The Training and Visit (T & V) System has been one of the most significant extension organizational developments in the last decades. Billions of dollars have been invested in this system by the World Bank since 1975. The system has been diffused very rapidly, first in South and South East Asia where it has been shown to increase the effectiveness of agricultural extension in irrigated areas in a number of countries, thus contributing to rapid increases in food production, and later in Africa where it was not always so successful.

The system tries to achieve changes in production technologies used by the majority of farmers through assistance from well trained extension agents who have close links with agricultural research (10).

10.9.1 Aims

Management has four main tasks:

(1) to develop the basic framework of the extension system in which everybody knows what he is supposed to do;
(2) to organize the support necessary to enable all extension agents to do their work well;
(3) to supervise how well extension agents perform their task and, if necessary, to help them perform this task better;
(4) to coordinate the extension work with agencies outside the extension service, such as research, provision of supplies and marketing.

10.9.2 Organization

The T & V System has an hierarchical organization with one extension agent (VEW) for about 800 farmers. In turn, eight extension agents are supervised by an Agricultural Extension Officer (AEO), and eight AEOs are supervised by a Provincial Extension Officer who is assisted by three to five Subject Matter Specialists (SMS). In a large country there may also be several higher level supervisors, each in charge of about eight subordinates and assisted by SMSs.

It is physically impossible for extension agents to meet all their farmers regularly. Hence, about 80 contact farmers are selected and visited every two weeks on a fixed day, preferably in their fields where other farmers can attend and join the discussion or demonstration of improved practices. Each supervisor, even those responsible for an area with several million farmers, is supposed to spend at least half his or her time in the field to check on work progress and to identify problems to be solved by management. Paperwork does not increase agricultural production and hence should be kept to a minimum.

10.9.3 Key features

(1) Staff are professionals with a sound knowledge of agricultural research, farmers' experience and factors limiting the productivity in agriculture.
(2) There is one extension service for all aspects of agriculture.
(3) The service is responsible only for extension. All other tasks required for agricultural and rural development, such as distribution of fertilizers and other inputs, marketing of products or provision of credit, are performed by other organizations or by private business. The extension service should cooperate closely with these organizations.
(4) There is a single line of technical and administrative command. Everybody knows who their supervisor is and what they are expected to do. The supervisor visits regularly to check their work and to help them do it better. The extension agents are supported by, but not supervised by, SMSs.

(5) The service concentrates its efforts on the main prospects for increasing agricultural production in the existing situation, that is, with the resources different groups of farmers have available.
(6) The service is oriented entirely to farmers' field problems. Everybody has to contribute to solving these problems, so there should be an efficient communication line up and down between the extension agent and the top of the organization.
(7) There is regular and continuous training of the whole staff to ensure every member performs well. There is two-way communication between extension and research in this training.
(8) Extension and research are closely linked to ensure agricultural research performs its main task of solving farmers' production problems.
(9) Research findings are tested in trials on farmers' fields before they are recommended to farmers. Farmers then are advised to test these recommendations further on a small scale.
(10) The quality and efficiency of the extension service is monitored and evaluated continuously to stimulate improvement.

10.9.4 The importance of professional staff

Promotion in the extension service is based mainly on staff members' performance. An effective extension agent or SMS can be promoted to a higher rank while continuing to work in the same position. These positions require capable people. It takes time for them to gain farmers' confidence, so they should not be transferred to another district very often.

Extension staff training receives substantial attention in the T & V System because well trained extension agents are the basis for an effective and professional extension service. In this training extension agents learn:

- to identify relevant production technology needed by farmers;
- diagnostic skills; and
- appropriate communication techniques.

Every two weeks the extension agents (VEWs) receive one day of training by the SMSs. This training should be based on analyses of the extension agents' experiences in the past two weeks. Most of this training should be given in the field where the extension agents can perform the tasks they later will have to teach themselves. Unfortunately, this training sometimes is given as straight classroom lectures based on notes taken by the SMS when a university student.

Every month the SMSs meet with research workers to discuss agricultural production problems for the next month. These meetings also serve as training for the SMSs. There also are regular short courses on production technology and extension methodology during which the extension agents are released from their normal duties. Some extension agents should be given opportunities to participate in longer term training towards degrees in agriculture.

The extension service has an important duty to inform research organizations of farmers' problems which require solving. It is pointless doing research on the assumption that irrigation and resources to buy fertilizers and other inputs are available to farmers who in fact have few resources and rely on natural rainfall. Under these circumstances research planners should focus on optimal use of scarce resources if they wish to make a major contribution to agricultural development.

10.9.5 Difficulties of the T & V System

We touch here on one of the difficulties of the T & V System. It assumes that everybody is working for agricultural development, whereas in fact some staff members of the extension service and cooperating organizations have other goals as well. For example, research workers may be more interested in developing technology which leads to the highest yields rather than to optimal yields with limited resources. Publication of these research findings in scientific journals may give them more opportunity for promotion and more status than cooperation with local extension agents to solve farmers' problems. Some research workers tend to look down on the less well educated extension agents who work for farmers, the low status group in many societies.

There are also other reasons why the T & V System is not always implemented in the way the theory mentioned above says it should be. In reality the tradition of a top-down extension approach is often continued in the T & V System and the crucial key feature (6) is not achieved. It is not a very serious problem in irrigated areas where there are technologies on the shelf which are not yet widely used, but which are profitable for the farmers as well as for the country. This was the situation in many Asian countries when the system was introduced shortly after the High Yielding Varieties became available. There the extension agents could work more or less as postmen, transferring messages they received from SMSs and researchers without adjusting them to the specific needs of each of their farmers who work in a rather uniform agro-ecological and socio-economic situation. Location-specific messages are needed in most of the rainfed areas. Therefore a much more participatory approach is required. This has been achieved much less often in government extension services applying the T & V System than in NGOs working in a more participatory tradition. In section 9.5 we pointed out how Hayward has shown that a participatory extension approach is necessary in these situations.

The T & V System also assumes that other agencies or private firms organize an effective supply of inputs and marketing of products. In remote areas this often does not happen until distribution of inputs by the extension service has first created a demand for these inputs and produced enough products to make it profitable for a private firm to come and buy them.

After their World Bank loan for the T & V System expired many governments were not able or willing to continue to finance this system. It requires rather high expenditure on transport for regular training and farm visits. In the first year of implementation it was possible to give the contact farmers valuable new information every fortnight, but after a few years the extension agents may have to repeat what they told them the previous year. This is not perceived to be useful by these farmers and hence they do not press the government to continue financing the system.

The system assumed that the contact farmer would be willing to act as an unpaid extension agent for colleagues. There have been situations where this worked because the right opinion leaders had been selected with the help of the community. However, the contact farmers often were not willing to invest time and energy in performing this role. In several countries this situation has prompted the extension service to work with groups of farmers rather than with individual contact farmers.

Most South and South-East Asian countries have ceased using the T & V System. Indonesia has switched to the Farmer Field Schools (which we discussed in Section 5.3.7). India still uses T & V but serious proposals have been made there to switch to a more participatory approach (11). The government prefers to pay more attention to poor farmers in rainfed areas in order to prevent a large scale migration to the cities. In irrigated areas there is also more need now for location specific solutions to make optimal use of irrigation water and fertilizers, and it is considered advisable to introduce IPM.

Some staff members of the World Bank itself are convinced the T & V System should be abandoned completely (12).

Despite these criticisms, we have discussed the T & V System in detail for two reasons:

(1) It is likely that many of our readers will have or will seek a job in a T & V extension system.
(2) People who criticize the T & V System are not always very clear about the alternative management system they would prefer, keeping in mind the limited skills of available extension staff. Many of the NGOs do a good job on a small scale, but it is a problem of different magnitude to change the management system of existing extension organizations, some of which have thousands of staff members serving millions of farmers on very limited budgets. It is not yet clear how that can be done.

10.10 Need for research

Research into the structure and functions of organizations is increasing rapidly. Before the Second World War most research of this type was conducted in factories. Today the emphasis has changed to include hospitals, research institutes and other service organizations, as well as commercial firms. Unfortunately, little attention has been paid to extension organizations, despite the fact that poor organization of those services often creates a bottleneck for their delivery. Hence it is difficult to specify how they should be improved.

Among the many questions requiring answers are:

(1) In what way does the optimal structure of the organization depend on the task of the extension service?
(2) What is the best way of organizing internal communication in the organization? What is the best way of setting up a management information system which quickly can give trustworthy information about the way extension agents work, the difficulties they encounter and the results they achieve?
(3) Which tasks can be combined in an extension organization and which cannot?
(4) To what degree is specialization desirable? How can the expert contribution of specialist knowledge be combined with optimal coordination of the work?
(5) To what degree, in what way and by whom are the managerial tasks fulfilled which were discussed in Section 10.2? What constraints prevent optimal completion of these tasks?
(6) How can different organizations serving the same target group cooperate in an organizational network?

Relatively little research has been published when compared with the large amounts of money which have been invested in the T & V System. Ideas published by different World Bank officers on the optimal way to organize agricultural extension are often based on systematic thinking about their field experiences, sometimes supported by (economic) theories. Their conclusions would be more convincing if based on empirical research, as is shown by the ISNAR studies on the organization and management of on-farm client-oriented research, and on research-technology transfer linkages (13). External evaluations would raise some different questions from those produced by internal evaluations, as we discussed in Chapter 8. Little money has been made available to universities to conduct external evaluations, mainly because universities are considered to be too theoretical in their approach. This attitude might be changed if selected university staff members were involved in action oriented research. Improvement of the pre-service training of extension staff could be a side effect from this action. Major research questions include ways of increasing the effectiveness of contact farmers, and the factors which influence the skills and motivations of extension agents.

10.11 Chapter summary

An extension organization requires:

- effective communication to management and to research of the problems faced by extension agents;
- effective communication of research findings to the extension agents;
- effective communication of changes in the environment to all concerned;
- a high level of motivation among all extension agents to work towards organizational goals; and
- sufficient flexibility which allows the organization to adjust rapidly to changing circumstances.

Communication and motivation can be improved by using a leadership style which gives extension agents the opportunity to participate in decisions regarding organizational goals and the ways these goals can be achieved. This enables management to be well informed about the extent to which the extension agents perform their roles as they are expected to do, as well as the reasons why they deviate from these expectations. The optimal style of leadership depends on the culture of the country, and the expectations and education level of the extension agents.

Management is responsible for ensuring that:

- clear decisions are taken regarding goals and target groups;
- the organization has extension agents who are capable and willing to achieve these goals;
- they have the resources required to achieve these goals;
- they know what each is expected to do; and
- they have the information required to perform this task.

It is desirable that normal contact with farmers is carried out by extension agents who have a good general knowledge of their clients' problems. They should be supported by specialists who are well informed about research findings related to their discipline or branch.

Female extension agents can play a valuable role in an agricultural extension organization, especially because of their ability to communicate more readily with female farmers and farm women. Their exact role differs according to the culture of the country.

The agricultural extension service usually is one of the divisions of the Ministry or Department of Agriculture, but extension also may be one of the functions of a university, a farmers' union, a board for the development of a settlement or an irrigation area or a certain crop, a commercial company or a private consultant. Each setting has its advantages and disadvantages. In this chapter we discussed the advantages and disadvantages of privatization of government extension services.

The T & V System is an important management system which concentrates the attention of the whole extension organization on problems farmers face in their fields. The main task of management is to enable the extension agents to help farmers effectively. Neither this system nor any other is the optimal management system in all situations.

At present a major task of extension managers is to manage a process of change in their organisation. All over the world we see a change, from the transfer of new technologies or providing farmers with solutions for their problems, towards facilitating and guiding a process in which farmers and extension agents jointly develop these solutions and learn from their experience. This change is necessary in a rapidly changing environment. However, this can only be realised through a change in the structure and the culture of the organisation which make it more flexible and participatory.

☞ Discussion questions

1 Imagine you must select someone for the position of extension agent with the aid of an interview, references and a trial period. Which characteristics would you attach most importance to (knowledge, insight, attitude, skill, etc.)? Why? How are these dependent on the position the extension agent is going to fill in the extension organization?

2 Much of the effectiveness of an extension organization depends on its extension agents' motivation to keep in close contact with farmers to help solve their problems. What can the head of a regional extension service with 30 extension agents do to strengthen this motivation?

3 Direction of extension requires simultaneously decentralization of decision-making and coordination. Do you agree with this statement? Why? How can you best achieve your own ideas about this situation?

4 How do you think that the governmental extension service in your country should and could be adjusted to its changing environment?

5 What is the best way to reach female farmers and farmers' wives with agricultural extension in your country?

6 In Section 10.3 we discussed six factors which should influence the choice of leadership style in an extension organization. Which style do you consider desirable in the agricultural extension organization of your country? Why?

Guide to further reading

Antholt, C.H. (1994) *Getting ready for the twenty-first century; technical change and institutional modernization in agriculture.* World Bank Technical Paper, 217, Washington DC.

Benor, D. and Baxter, M. (1984) *Training and Visit Extension.* World Bank, Washington DC.

Blum, A. and Isaac, M. (1990) Adapting the Training and Visit System to changing socio-cultural and agro-ecological conditions. *Journal of Extension Systems*, **6**, 45–66.

Buford, J. A., Bedian, A. G. and Lindner, J. R. (1995) *Management in Extension.* Ohio State University Press, Columbus.

Cary, J.W. and Rivera, W.M. (1996) Privatising Agricultural Extension. Chapter 24 in: *Improving Agricultural Extension* (ed. B.E. Swanson). FAO, Rome.

Claar, J.B. and Bentz, R.P. (1984) Organizational design and extension administration. Chapter 12 in *Agricultural Extension* (ed. B.E. Swanson). FAO, Rome.

Frohman, M.A. and Havelock, R.G. (1969) The organizational context of dissemination and utilization. Chapter 6 in: *Planning for Innovation Through the Dissemination and Utilization of Knowledge* (R.G. Havelock *et al.*) Center for Research on the Utilization of Scientific Knowledge, University of Michigan, Ann Arbor.

Hellriegel, D., Slocum, J.W. and Woodman, R.W. (1992) *Organizational Behavior*, 6th edn. West Publishing, St Paul.

Leonard, D.K. (1977) *Reaching the Peasant Farmer: Organization Theory and Practice in Kenya.* University of Chicago Press, Chicago.

Moris, J. (1991) *Extension Alternatives in Tropical Africa.* Overseas Development Institute, London.

Pareek, U. and Venkateswara Rao, T. (1992) *Designing and Managing Human Resource Systems*, 2nd edn. Oxford and IBH (Indian Book House), New Delhi.

Smith, P. (1989) *Management in Agricultural and Rural Development.* Elsevier, London.

Umali, D.L. and Schwartz, L. (1994) *Public and private agricultural extension: beyond traditional frontiers.* World Bank Discussion Paper, 236, Washington DC.

11 The Role of Agricultural Extension

We have touched many times in previous chapters on the role of agricultural extension. In this chapter we will discuss first the role of agricultural extension in changing our society, and then the role of the agricultural extension agent in this process. In this way we are, in fact, summarizing major parts of this book.

11.1 Changing our society

A major goal of the agricultural development policy in most countries is to increase food production at a similar rate to that at which the demand for food is increasing, and at a cost which is competitive on world markets. It is appreciated more and more that such development must be sustainable, and that often it must be done in a different way than it was in the past.

An effective agricultural extension organization is critically important in this situation, especially in less industrialized countries. There are also many problems which decrease the effectiveness of these organizations in some, but not in all, countries. Problems include the following:

(1) Appropriate technology is not available to extend to the farmers.
(2) There are no effective linkages between extension organizations and agricultural research institutions.
(3) Field level personnel lack practical training in agricultural technology.
(4) Extension personnel lack training in extension methods and communication skills.
(5) Extension personnel lack adequate transport facilities (i.e. mobility) to reach farmers effectively.
(6) Extension personnel lack essential teaching aids, demonstration materials and communication equipment.

(7) Due to organizational problems, extension personnel have many other tasks besides extension work.

The role of an agricultural extension agent is to help farmers form sound opinions and to make good decisions by communicating with them and providing them with information they need. Farmers' opinions and decisions are based on their image of the reality in which they live and on their expectations of the consequences of their actions in this reality. However, these expectations are not always correct because their image of reality never agrees completely with reality itself. The extension agent therefore has a major task in helping farmers come to terms with reality. This gives farmers more control over their own lives because their actions then are likely to have the desired consequences more frequently. Thus, by achieving desired consequences more frequently, the farmers would be better adapted to take control of their own lives.

The major role of extension in many countries in the past was seen to be transfer of new technologies from researchers to the farmers. Now it is seen more as a process of helping farmers to make their own decisions by increasing the range of options from which they can choose, and by helping them to develop insight into the consequences of each option.

Farmers seek information not only from their extension agent but also from a range of sources, including their own experiences and those of their colleagues to develop this insight. Government policies, such as those on environmental problems, have an increasing impact on the options open to farmers.

The opinions and decisions also are based on farmers' values, although, as they are not always clear about this relationship, extension agents also must help them clarify it. Hence the agents can help farmers with their decision-making on their pathway towards knowledge as well as on their pathway towards choice. The pathway towards knowledge generally receives most attention from the agents, but the information they provide from this pathway is effective only if the farmers realize that it helps them on their pathway towards choice.

People acquire their images of the reality in which they live by:

- learning from their own experiences;
- by observing other people's experiences;
- by talking with other people about their experiences and about research findings; and
- by thinking about information they have gained in these ways.

The extension agents' role is to promote and supplement this learning process. In doing so they will improve their own image of reality by learning from the farmers.

11.2 The role of extension agents

Farmers have expectations about the way extension agents will help them, but the agents' superiors also have expectations about the agents' role. The agent, as person in the middle, can be in trouble if the role expectations of these two groups conflict. There is a high probability this will occur if:

(1) The superiors expect the agents to implement an agricultural development programme which is not in the farmers' best interests, even though it may be in the national interest, such as increasing export earnings, for example.
(2) Farmers expect their extension agents to provide services rather than to help them with their education, especially if the agents' role has not been explained to them.
(3) Extension agents are expected not only to perform an extension role, but also some other role, such as policing regulations or supervising credit, which conflicts with the extension role.

Several studies show that extension agents in industrialized countries are more interested in the farmers' opinions of their work than those of their superiors, mainly because they have to work with these farmers every day. They will try if possible to please their superiors as well as the farmers, but if this is impossible they will often place higher priority on the farmers' wishes. Extension agents in the majority of less industrialized countries probably are more interested in their superiors' opinions of their worth. Some reasons for this difference might be:

- farmers in industrialized countries have higher status in their society;
- the fact that many extension agents in industrialized countries would have preferred to have been farmers themselves, whereas many farmers in less industrialized countries would like their children to find government jobs;
- the less hierarchical structure of the society in many industrialized countries; and
- the fear of extension agents in less industrialized countries of losing their jobs or of missing a promotion which might make it difficult for them to continue supporting their families.

An Australian study showed that farmers prefer information sources which have a practical approach, which can show the results of recommendations, and which have considerable local knowledge as well as knowledge of the economic consequences of the recommendations (1). It was much less important for the source to be up-to-date, technically reliable and able to cite experimental results. These Australian farmers also considered it important for the source to be unbiased, honest and trustworthy. We have the impression that Australian farmers are no different in these respects from farmers in other countries. However, not all agricultural extension agents can meet these preferred criteria.

11.3 Fulfilling these roles

The extension organization and the extension agent should keep the following points in mind when helping farmers to form sound opinions and to make effective decisions:

(1) The extension manager and agent must clarify in their own minds the circumstances under which they may influence farmers and those under which they *must* influence them. It is far too simple a view to say the manager and the agent may and must help farmers to achieve their goals in the best way possible, because:

- (a) farmers can harm others in achieving their goals;
- (b) different farmers may have conflicting goals, so the agent cannot help to achieve all these goals harmoniously;
- (c) this view is based on the following assumptions:
  - it is always possible to help farmers to weigh up conflicting goals against each other;
  - farmers are prepared and in a position to choose between these goals if it appears they cannot all be achieved at the same time;
  - the consequences of their chosen solution can be predicted efficiently.

  These assumptions are not always justified.
- (d) the extension organization may restrict the number of possible solutions an agent can offer to farmers for their problems;
- (e) agents' and farmers' goals may conflict with each other.

(2) Farmers' trust in their agents is an essential condition for good extension. In order to win this trust farmers must be convinced that agents are trying to serve their interests, that they can empathise with the agents and that they are experts in their field. Agents will be more likely to win this trust if they visit farmers in their fields or at their house, rather than expecting farmers to visit their office. Agents who work with farmers in their familiar home environment can demonstrate that they are really interested in their problems and that they have sufficient expertise to help them solve their problems. Naturally this approach makes extension more expensive.

(3) An individual's actions are constrained by his environment. The extension manager and agent should ask themselves if they should help farmers make the most of existing opportunities in their environment or if they can expect to achieve more by helping farmers to influence the environment itself. What will agents do if they are convinced that extension will be effective only in a changed environment if they are not in a position to bring about this change?

(4) Extension agents who wish to help farmers must try to see everything from the farmers' point of view, their problems, goals, knowledge and their use of language. It may be useful to help farmers express the way in which their feelings influence their behaviour so these can be discussed openly. There is no point in giving a scientifically correct solution if farmers are not yet aware of the problem. The agents will be in a better position to make farmers aware of their problems if they have a high degree of empathy for the way farmers think. They also may be in a better position to make farmers aware of the need for change. Extension information will be effective only if it fits into the farmers' decision-making process and is compatible with their way of thinking and of using the language. Hence it is more important for an extension agent to be a good listener than a good speaker.

(5) It is much better for farmers to find their own solutions to a problem than for an extension agent to find it for them. Farmers will be more motivated to implement their own solutions in practice, and will feel more responsible for their own decisions. Furthermore, if their own solution appears to be successful they will gain self-confidence and will have learned something useful about problem solving.

(6) Everyone's behaviour is strongly influenced by positive reinforcement of their own past experiences, as well as by the norms of the group they belong to or would like to join. Farmers also will be more likely to change their behaviour if they discover for themselves that their knowledge and insight gained from past experience is no longer adequate to deal with current problems. They also are more likely to integrate this new information into their cognitive map if they work it out for themselves rather than being told what to do by an extension agent.

It is often easier to change the norms of a group as a whole rather than persuade an individual to deviate from these norms, and in much the same way as an individual, the group will have to discover for itself that such a change is desirable if it is to change its norms.

(7) All members of a group are unlikely to adopt innovations simultaneously. Less progressive group members are more likely to be influenced indirectly by their fellow members who are opinion leaders than directly by an extension agent.

(8) Effective communication is extremely difficult without feedback about how the receiver interprets the source's message. Pre-testing mass communication messages can give useful information on which to base message changes if necessary. An extension agent must pay close attention to audience reactions during group discussions and lectures, as one-way communication usually has little effect.

(9) Extension managers and agents can make systematic use of information

gained from evaluation of extension programmes, how they have been implemented, their effects and the reasons why the results are as they are. Feedback of this kind is essential for effective extension work.

(10) An agricultural extension agent should understand many aspects of:
- crop and livestock production
- farming as a business
- agricultural development processes
- farmers and the way they learn
- rural society.

Agricultural extension agents can fulfil the roles and tasks outlined above if they satisfy certain requirements themselves. They must have adequate technical knowledge to solve farmers' problems, or they must be able to obtain this knowledge when required. Their information must be accurate. Farmers quickly will lose their confidence in an agent who gives them incorrect advice, especially if it had been possible to give correct information.

Agents also must develop relationships with farmers that are favourable for their development. Amongst other things, agents should be aware of how their personal feelings may influence their relationships with farmers.

Many extension organizations have a narrower view of extension and the extension agent's role than we have outlined in this chapter. They see it as a process of supplying information to farmers on demand, and of introducing technical changes in agriculture which they consider to be desirable, rather than one of promoting farmers' development and independence. This type of role interpretation requires a lower standard of performance by extension agents because the changes they try to achieve are more limited.

11.4 The role of extension administrators

Administrators of an extension organization also should consider these ten points in order to manage their organization in such a way that their agents can work effectively with their farmers. They must also pay attention to some other points:

(1) Which role can and should their extension programme play in implementing the agricultural development policies of their government? Which of these policy goals can be achieved solely through communication and education, and which through communication and education combined with other policy instruments?

(2) How can effective communication be implemented between agricultural extension and research?

(3) How can administrators manage the planning of an extension programme in which decisions are taken about the following issues:

(a) the goals and the nature of the changes the organization wishes to achieve;
(b) the target group;
(c) the contents of extension messages by which the organization hopes to achieve these changes;
(d) the communication methods and channels the organization will use and how they will be used;
(e) the structures, persons and organizations with which the organization expects to achieve its goals. Who is going to do what and how, and is each in tune with the other? How does the organization cooperate or compete with other organizations?
(f) the time available for planning and implementing the extension programme?

(4) How can farmers participate in planning extension programmes, and in development and testing of extension messages?

(5) How can the administrator stimulate the whole organization to increase its effectiveness by learning from its experience?

(6) How can the administrator manage the extension organization in such a way that:

(a) it is able to attract and retain capable staff members;
(b) all staff members have a clear picture of the goals of the extension organization and how they can contribute most effectively to achieving these goals;
(c) all staff members are well trained in agricultural technology and extension education to ensure they can perform their tasks effectively;
(d) all staff members are motivated to perform their tasks effectively and to win the farmers' trust;
(e) the promotion of staff members is based mainly on merit;
(f) there is effective communication upwards, downwards and laterally within the extension organization; and
(g) there is good cooperation between the extension organization and other organizations which play a role in agricultural development?

(7) How can the administrator ensure that sufficient resources are available to perform the task faced by the extension organization in implementing agricultural development policy, and in providing the information and education farmers need? What situation will the organization try to reach in the market for information and education? We predicted much change in many extension organizations in the near future. This makes it challenging for people, who feel they may have to contribute to this process of change, to

work in agricultural extension. An extension organization may not be the right place any longer for people who prefer a secure and stable job.

11.5 Chapter summary

The extension agent's role is to help farmers form sound opinions and make effective decisions. Farmers are encouraged to develop a high degree of independence in their decision-making. Many factors which contribute to the extension agent's role are summarized in this chapter.

☞ Discussion questions

1 How can an extension agent stimulate personal development of farmers? How can the agent hinder this development?
2 An extension agent should try to achieve good relationships with superiors as well as with farmers. Under which conditions do you expect an agent to achieve both these aims simultaneously, and when it is difficult or impossible to achieve them simultaneously?
3 How can an agricultural extension agent or manager increase farmers' trust and confidence in him or her or the extension organization?
4 Our behaviour is influenced to a large degree by our insights and knowledge that we have found through previous experience to be correct. Imagine that an extension agent is convinced that farmers' insights are no longer correct. How can the agent help them revise these insights?

Guide to further reading

This chapter does not outline all possible approaches to an extension agent's role. Somewhat different approaches are presented in:

Albrecht, H., Bergmann, H., Diederich, G., Grosser, E., Hoffman, V., Keller, P., Payr., G. and Sülzer, R. (1990) *Agricultural Extension.* Deutsche Gesellschaft für Technische Zusammenarbeit (GTZ), Eschborn.

Benor, D. and Baxter, M. (1984) *Training and Visit Extension.* World Bank, Washington DC.

Bollinger, E., Reinhard, P. and Zellweger, T. (1992) *Agricultural Extension: Guidelines for Extension Workers in Rural Areas.* Skat, St. Gallen, Switzerland.

Christoplos, I. and Nitsch, U. (1996) *Pluralism and the Extension Agent: Changing Concepts and Approaches in Rural Extension.* SIDA, Stockholm.

Röling, N. (1988) *Extension Science: Information Systems in Agricultural Development.* Cambridge University Press, Cambridge.

Swanson, B.E. (ed.) (1990) *Global Consultation on Agricultural Extension.* FAO, Rome.

Glossary

In this list we attempt to define or describe concepts which occur in this book and are considered important. The list is meant to be a help for the student. However, for reasons of space the reader must appreciate that the definitions and descriptions will explain only certain aspects of the concept. Besides this, many concepts in the social sciences are used in different ways because of the different schools of thought within these sciences. Generally the description chosen is the most crucial, or related to the way the concept is used in this book.

Action research A form of social research which aims at better insight into the problem by learning from the experience gained in an attempt to solve this problem.

Actor A person who plays an active role in the Agricultural Knowledge and Information System (AKIS).

Adopter categories Members of a social unit classified on the basis of the speed with which they adopt innovations of a certain type, e.g. innovations based on agricultural research.

Adoption (of innovations) Decision to apply an innovation and to continue to use it.

Adoption process (with regard to an innovation) The changes that take place within individuals with regard to an innovation from the moment that they first become aware of the innovation to the final decision to use it or not.

Adoption research Research into the way that members of a social system adopt innovations. See also *Diffusion research.*

Agricultural development Change in agricultural production techniques and in farming systems towards a more desirable situation, usually one in which farmers use more agricultural research findings and in which there is less subsistence and more market-oriented agriculture.

Agricultural Knowledge and Information System (AKIS) The persons, networks and institutions, and the interfaces between them, which engage in or manage the generation, diffusion and utilization of knowledge and information, and which potentially work synergistically to improve the goodness of fit between knowledge and environment, and the technology used in agriculture.

Anthropology A science which studies man, including his physical characteristics and his culture. We can distinguish between physical and cultural anthropology. See *Culture*.

Applied science The scientific activities which apply the principles and insights from one or more sciences to the analysis and understanding of a concrete phenomenon in order to find a suitable solution to a problem.

Aspiration level The standards set by individuals for themselves regarding the level they want to achieve with future performances. Their experiences in relation to success, the difficulty of the tasks, the norms in the social environment and personal factors are important here.

Attitude The more or less permanent feelings, thoughts and predispositions people have about certain aspects of their environment. Three components are recognized: knowledge, feelings and inclination to act.

Audio-visual aids Aids to support educational and communication processes, which are perceived by seeing and by hearing.

Authoritarian leadership A form of giving leadership in which the leader(s) fix the goals and behaviour code of a group. The leader does not have to take into account group members' opinions about the decisions. The leaders expect obedience from the group members.

Bureaucracy A form of organization characterized by central leadership, hierarchical regulation of functionaries, decision process according to general rules and routine procedures, impersonal relations between functionaries and the general public.

CD-ROM A Compact Disc (Read Only Memory) which contains large amounts of digital and/or graphic information.

Change agent A person who tries to stimulate change among people or organizations. An extension agent is an example of a change agent.

Client (in extension education) A general indication of the individual to whom extension activity is directed.

Cognition The mental activity or the combination of mental activities, resulting in the knowledge or being aware of something. Also something one knows of, is aware of, that one knows.

Cognitive map The idea that someone has formed of reality or parts thereof.

Communication (noun) (1) The process which takes place if people or groups communicate with each other. (2) The act of communication. (3) The way of communicating.

Communication (verb) The process of sending and receiving messages through channels which establishes common meanings between a source and a receiver.

Communication channel The way in which people or groups are in contact with each other, which makes communication possible, e.g. the radio.

Communication process The process that takes place when people or groups communicate with each other. When describing this process the accent can be placed on its elements, stages that take place in the communication, participation activities of those involved, and rules that are taken into account by those involved.

Counselling A form of mutual discussion help in which the expert's function is to help clarify the relationship of the client with him or herself and with others.

Culture A complex whole of norms, values, knowledge, ideas, art, science, laws, habits and other abilities acquired by a person as the member of a society.

Database An electronic information storage system which may contain social, technical or bibliographic (i.e. library) information which can be searched and retrieved rapidly using appropriate software and computer.

Decision-making All the considerations which play a role in making a choice from alternative action possibilities. See *Decision-making process*.

Decision-making process The way in which decision-making takes place. When describing a decision-making process different emphases can be placed on the stages that one passes through, on the degree of rationality, on the relative participation of those involved in (aspects of) the decision-making, and on the rules that are followed.

Decision support systems Simulation models used to calculate feed rations, pesticide applications and other on-farm calculations and advisory activities.

Decoding The process by which a receiver transforms signals in messages into cognitions. See *Encoding*.

Diagnosis-prescription A way of acting which the helper can use on a client. The helper diagnoses the nature of the problem and gives a solution.

Dialogue A discussion between an extension agent and a single client to help the client to define a problem and to find a satisfactory solution.

Diffusion research Research into the way which innovations (generally) spread among the members of a social system. See also *Adoption research*, with which it is usually combined.

Directive Attempting to change the behaviour of others through discussion, by contributing one's own vision without paying attention to the insights of the other(s), or on the assumption that one knows what the insights of the other(s) are.

Education Usually a process of learning within formally structured social institutions, which are organized for this purpose, e.g. schools. Some use a wider definition and include non-formal education or learning outside these institutions, e.g. from news programmes on the radio. *Extension education* refers to this wider definition.

Empathy The ability to feel another's feelings in the way the other person does, and the ability to communicate that similarity in feeling.

Encoding The transformation of messages into signals.

Ethic Branch of philosophy which is dealing with what is good or bad, and right and wrong, and with moral duties and obligations.

Evaluation A policy management instrument which collects and analyses information so that the relevance, effect and consequences of activities are determined as systematically and objectively as possible in order to improve present and future activities, such as planning, programming, decision-making, and programme execution to comply with policy goals.

Evaluation, formative Tries to increase the effectiveness of an extension campaign by collecting information on reactions of the target group and experiences of the extension agents during implementation of the programme.

Evaluation, summative Tries to measure the changes achieved with a completed extension programme in order to use this experience to improve other extension programmes.

Evaluation research The use of scientific methods in evaluation.

Expert systems An important development in information technology used to advise the farmer which alternative to choose from a wide range of possible alternatives by processing data from a large number of variables according to decision rules which are normally used by experts.

Extension A form of conscious social influence. The conscious communication of information to help people form sound opinions and make good decisions.

Extension administrator A person whose main task is administration of an extension organization or a unit of such an organization, usually at a senior level. Concerned with policy making and implementation.

Extension agent/officer A person whose main task is to give extension assistance or to manage an extension organization at the field level.

Extension education A science in which strategic questions associated with the extension process are studied. Extension education tries to bridge the gap between the social sciences and the practice of extension.

Extension manager A person whose main task is management of field extension agents and resources. Usually a regional or district position.

Extension methods The methods of communication which can be used in extension for influencing the target groups.

Extension organization/service The name given to the organization which directs itself mainly at planning, execution and evaluation of the activities of extension agents. This organization also is considered to be a (desirable) link between the developers and the users of new knowledge and insights.

Extension programme A plan of the goals and target groups an extension organization or extension unit tries to reach, and of the ways it will try to reach them.

Farming Systems Research (FSR) A type of research through which an interdisciplinary research team tries to gain as complete knowledge as possible of the existing farming system in order to assess whether a new technology helps farmers in achieving their goals under those environmental conditions.

Feedback The process in which knowledge of the surroundings or of the consequences of actions of a system lead to adjustments of future actions, seen in the light of achieving a certain goal.

Flannel board A display board covered with flannel to which figures and words can be stuck if they are backed with sandpaper or other suitable material.

Flipchart A visual aid, namely a frame on which a pack of large sheets of paper is attached. The sheets are turned over after they have been written on, as with a writing pad, thus allowing you to return to what was written up earlier.

Folk media Communication media which do not use modern technology, but are based on traditional communication methods, such as itinerant singers, drama or puppet shows.

Frame of reference The whole of values, norms, convictions and assumptions on the basis of which someone judges or acts. The frame of reference of individuals is influenced by the group to which they belong or the group(s) or the social group to which they would like to belong.

Generalist A person who collects knowledge in various (specialized) areas in order to apply this knowledge.

Generalization The formulation of general concepts based on a number of specific observations. Note: many generalizations in the social sciences are bound by time or place.

Goal The end towards which actions, e.g. extension programmes, are moving.

Group Used in the sense of a social group: a collection of individuals related to each other by some common purpose and with some structure among them. Note: use of this concept is not restricted to the above sociological description.

Health education The conscious use of communication for stimulating people to behave in such a way as to maintain good health or to recover as rapidly as possible from a disease. Used in schools as well as in adult education programmes.

Holistic Studying a system (farm, farming system, organization) as a whole, because it cannot be understood properly by only studying its elements without understanding how these elements are interrelated.

Index (1) Scale. (2) The number on the scale given to the variable involved.

Indigenous knowledge That knowledge held in the rural society, usually based on the experience of many generations and unique to each cultural group. Usually it contains more information on local diversity and complexity than scientifically derived knowledge.

Individual-blame hypothesis The hypothesis that the low level of adoption of modern technology by some people is caused by their values and characteristics, e.g. their level of education.

Information The pattern people impose on the phenomena they are able to observe. This implies an interpretation of these phenomena to make them useful for reaching a certain goal by reducing the uncertainty in the consequences of actions.

Information technology Electronic systems for storing, retrieving, transmitting, receiving and processing information. Closely associated with recent developments with computers and telecommunications.

Innovation Ideas, methods or objects that are considered new for the individual.

Integrated Pest Management (IPM) An ecologically based pest control strategy which relies heavily on natural mortality factors such as natural enemies and weather, and seeks out control techniques which disrupt these factors as little as possible.

Intermediate goal Indicating a goal that someone or an organization strives for because it is recognized that achieving this is a condition for reaching the ultimate goal.

Knowledge To know about; contains an element of the concept insight. Knowledge is to be considered the vision of an explanation for the world in which we live, and knowledge is relative in the sense that the vision can differ between people and amongst others because of differences in experience. A distinction is made between everyday and scientific knowledge or between technical and social knowledge.

Knowledge transfer The process by which attempts are made to pass knowledge from one person or institution to others. When describing the transfer of knowledge we can note different dimensions, including the difference between participants, the method used, the context within which knowledge transfer takes place, the reason behind it and the effects.

Leadership The directing, influencing and controlling of others in pursuit of a group goal.

Learning Acquiring or improving the ability to perform a behavioural pattern through experience and practice.

Linkage A person who or an organizational unit which promotes two-way communication between two sub-systems, for example, between research and extension.

Listening An active process of receiving aural stimuli consisting of the five phases: receiving, understanding, remembering, evaluating and responding.

Management The executive function of planning, organizing, co-ordinating, directing, controlling and supervising an organization.

Manipulation The systematic attempt of individuals and groups to influence the behaviour, attitudes and opinions of other people without them being aware of it. This can be done by working on the unconscious desires and images of those to be influenced, by suggestion (psychological manipulation) or by changing or influencing the social surroundings (social manipulation). Also used in other publications as *influencing*.

Market research Research directed at gaining insight into (potential) markets for products. Developed in and for the production, price, advertising and distribution policy of commercial companies. Market research depends partly on insight into consumer behaviour. Market research bureaux work also for the non-profit sector.

Mass communication Public communication, in principle open to everyone. Used here as communication that takes place via the mass media.

Mass media Communication media which reach a large audience through printed matter or electronic methods, such as radio, television, films, newspapers, posters and books.

Meta communication Communication about communication, e.g. about the emotions extension agents arouse among their clients by the way they communicate with them.

Model A simplified, schematic reproduction of a pattern of discovered or suspected relationships between certain phenomena. A model always is created and described with a certain goal in mind and from a certain point of view. Therefore, only certain facets of the phenomenon are highlighted.

Modem An electronic device which connects a computer to a telephone line and converts between digital data in the computer and analogue signals on the telephone line.

Modern Used here for groups, individuals or organizations characterized by their positive as opposed to traditional attitude towards change. The term is related to development of the so-called modern society.

Monitoring A management technique in which extension agents collect data on the way an extension programme is implemented and the problems encountered by extension agents with this implementation. This information then can be used to solve such problems as quickly as possible.

Motivation The internal state which stimulates a person to carry out certain activities.

Multi-media approach To make use of a combination of a number of different communication media in an extension programme. This approach is based on the view that different media have different characteristics in terms of whom they reach, their effects, etc., and the idea that using complementary and/or overlapping media can be more effective.

Needs The condition in which a person experiences a lack of something and strives to overcome this lack. It is now accepted fairly generally that most needs, except purely physiological ones, arise from experience.

Network, electronic A set of computers linked by cables, telephone line or via satellite connections which enable transfer of information (data) between machines.

Network, personal A set of direct and indirect social relations, centred around given persons, which are instrumental to the achievement of the goals of these persons, and to the communication of their expectations, demands, needs, and aspirations with other members of the network.

Non-directive A method of discussion in which helpers do not give their own vision but, without pushing, stimulate the partners in discussion to express their thoughts and emotions about a certain subject. Some people are of the opinion that some degree of pushing can never be prevented.

Non-Governmental Organization (NGO) An organization which plays an intermediary role between the people and the government and/or tries to promote the welfare of a certain group of people. An NGO can be formed by members to improve their own situation, but also by outsiders who try to serve the interests of (usually poor) people. A more restrictive definition is also used: a non-profit voluntary organization engaged in the philanthropic pursuit of relief and development activities. We use the first definition.

Norm, social Rules of behaviour. Ways of behaving in certain situations, or ways we are expected to behave. A norm can apply to a society (general norms), for certain groups or for a small group (such as a family, or school). In the second case it is called a group norm. See *Social control.*

Opinion formation The process which leads to judgement(s) of an individual over something. Important elements of this process can be the (new) experience of individuals, the (new) knowledge that they have acquired, the social group(s) to which they belong and the social control of their behaviour and values.

Opinion leader A person who has a relatively large influence on the opinions of others in the group to which he or she belongs. Opinion leaders are seen as important contributors to the formation of public opinion about new ideas, situations, etc.

Organization Used here as a formal organization: a formal group with an explicit goal, set rules and procedures and a clear division of tasks with set rights and duties. Every organization also has an informal structure: the network of personal relations which arises spontaneously if individuals interact in a formal organization.

Organization development (OD) A process by which an organization becomes more capable of achieving changing goals in a changing environment, usually with the assistance of a consultant.

Organization structure The network of interactions and relations between members of an organization, in which the members are seen as occupying the positions of actors of their roles. The division of labour and power is regulated via the organization structure.

Participation a way of policy making in organizations. Those for whom the decisions have consequences are given the opportunity by the administration to express their opinion about policy proposals and make their wishes clear (usually only orally). Note: this concept is also used in different ways.

Participatory leadership A leadership style in which workers in an organization contribute to the decision-making process regarding the goals and the methods to reach these goals, and share the responsibility for these decisions.

Participatory Rural Appraisal (PRA) An approach to Rapid Rural Appraisal (RRA) in which the rural people themselves play an important role in collecting, analysing and interpreting the data.

Pathways towards knowledge and towards choice Analytic distinction within a decision-making proess. This distinction is made in order to get a grip on the interplay between relevant knowledge and choice in the process of decision-making.

Perception The process by which we receive information or stimuli from our environment and transform it into psychological awareness.

Persuasive Convincing; influencing attitudes and behaviour in a direction considered desirable by the source.

Pilot project Trial programme; testing a certain change strategy in a real situation on a small scale, the goal being to learn the difficulties which will occur with the execution of this change and to evaluate if this change is workable. It then can be decided if the policy should be put into action on a large scale. A pilot project should be seen as an experiment in reality.

Planning The process of determining (1) organizational objectives, (2) developing premises about the environment in which they have to be accomplished, (3) selecting a course of action for accomplishing the objectives, (4) initiating activities necessary to translate plans into action, and (5) evaluating the outcome of that planning.

Policy A form of directed action which indicates as clearly as possible what one wants to achieve, how one wants to do it and how much time will be taken to achieve the set goals. Policy can be seen as an attempt at consciously influencing and controlling future situations.

Pre-test Trial before introduction, as with extension publications and aids. Some people from the target group are shown the drafts. Their reactions are noted, and decisions are made whether the publications and aids are readable, entertaining and/or easily and correctly understood.

Problem a situation in which a person who has a goal does not know how to reach that goal.

Problem solving model A model of the research utilization process which stresses the need to look for all relevant research findings to solve the problems of persons and organizations.

Programmed instruction An instructional method in which the subject matter is divided into small units, some of which require an answer from the learner. This answers shows if the learner has understood the subject. Learners get immediate information whether or not their answers are correct. This stimulates their activity or corrects a misunderstanding.

Propaganda A conscious, systematic and organized effort to manipulate the decisions, actions and ideas of a large number of people, or to influence them with regard to a controversial subject, in a direction which is desirable from the point of view of the insights or interests of the propagandist.

Psychology a science which studies the nature, functions and phenomena of the human mind.

Public relations A communication function of the management of an organization through which it tries to adapt, to alter or to maintain the organizational environment for the purpose of achieving organizational goals.

Random sample A sample in which all combinations of persons have an equal chance of being selected for investigation (as respondents).

Rapid Rural Appraisal (RRA) An approach to research for understanding aspects of the rural society in which an interdisciplinary research team tries to optimize trade-offs between cost of research and relevance, timeliness and beneficial use of information, to offset biases, to triangulate conclusions by combining different research methods, to learn from rural people and adjust the direction of their research on the basis of what they have learned during the research proess.

Rational (problem solving) Reaching a solution to a problem by the use of logical, systematic reasoning; not guided by emotions.

Respondents People who have answered questions by an interviewer for a social survey. They are the people from whom social researchers usually obtain most data required for their research.

Role The complex of expectations with regard to the behaviour of individuals in a certain social position; these expectations are present among the individuals with whom they have a certain relationship. Roles regulate the interaction between individuals; offer support. One person can fulfil different roles, for example, father and student.

Self efficacy The perception a person has of his or her ability to perform a task well.

Simulation A technique for understanding processes by imitating a situation which happens in reality or can happen.

Social change A general term for the total transformation which occurs in societies and their most important components; the social dynamics of society. There is much difference of opinion about the nature of the powers behind social changes. For example, there are different opinions about the influence of ideas as opposed to the influence of economic order as a cause of change.

Social control (1) In a broad sense it implies the whole regulating mechanism in society. Social control sees that human behaviour conforms to the society's orders and guarantees the continuity of this order by correcting deviating behaviour. The methods cover a wide range from bodily force to soft psychological pressure. (2) In a restricted sense it implies influencing (controlling) people's behaviour in order to maintain group norms.

Social interaction Interaction between persons; the process of interplay and reciprocal influence of the actions of various individuals or groups, usually combined with communication. Social interaction can take various forms, such as reciprocal and one-sided, co-operative and conflicting.

Sociology A science which studies the culture and structure of societies and their influence on changes in these societies.

Stakeholders People and organizations who can influence a certain issue or who are affected in any way by what is done and how it is done.

Status The place that a person or social category holds in a society or within a certain group in relation to that held by others.

Strategy A way to achieve clearly specified goals with a combination of means and in a certain time period. By anticipating we try to predict what the opponent(s), oneself and/or nature can do.

Structure, social The network of social interactions and permanent relations between members of a group or society as occupants of positions or players of a role.

Subject Matter Specialist (SMS) An extension agent who has studied one scientific discipline or one specific problem areas. Such an agent's help can be sought by generalist extension agents or by specialists in other areas.

Survey research A type of research in which respondents are asked to provide facts or opinions with the aid of a fixed schedule of questions – a questionnaire. The questionnaire can be administered orally or in written form.

Synergy 'Cooperation' among people who, following their own interests and seeking their own goals, produce benefits for their society.

System-blame hypothesis An hypothesis that the low level of adoption of modern technology by some people is caused by the structure of the society in which they live and by their place in this structure. This makes it impossible, unprofitable or difficult for them to adopt modern technology.

Target group A general indication of the more or less related group(s) which the extension agent or the extension organization wants to reach or change. The target group can refer to individuals or organizations.

Target group, intermediary A group of people or organization(s) at which the extension agent or organization directs its activities in the hope that the ultimate target group will be reached or influenced via this group.

Target group segmentation Mental splitting by the change agent of the heterogeneous groups into sub-groups which are more homogeneous with respect to the goals of the change agent and/or the communication channels they use.

Training and Visit System (T & V System) An extension management system in which the Village Extension Agent (VEA) regularly visits contact farmers after receiving training from the Subject Matter Specialists (SMSs). Promoted in many less industrialized countries by the World Bank.

Value An abstract, generalized behaviour principle to which the members of a group have a strong positive emotional attachment. It is also used as a measure for judging the desirability of specific actions or goals.

Variable A general indication in statistical research of a characteristic which occurs in a number of individuals, objects, groups, etc., and that can take on various values, for example, the age of an individual or the inhabitants of cities.

Village Extension Agent (or Workers) (VEA) A generalist extension agent at the local level, which might include several villages, responsible for the first contact of farmers with the extension organization.

Chapter Notes

Chapter 2

1 We are grateful to Mr Gwyn Jones of the Agricultural Extension and Rural Development Centre of the University of Reading for this information.
2 Birkhaeuser, D., Evenson, R.E. and Feder, G. (1991) The economic impact of agricultural extension. *Economic Development and Cultural Change*, **39**, 607–650.
3 Mosher, A.T. (1966) *Getting Agriculture Moving; Essentials for Development and Modernization.* Praeger, New York.
4 Chamala, S. and Keith, K. (1995) *Participative Approaches for Landcare: Perspectives, Policies, Programmes.* Department of Primary Industry, Brisbane.
5 This discussion is based on: Hoffmann, V. (1992) Beratungsansatze: Von der Uniform zum Maszanzug. In: *Beratung als Lebenshilfe; Humane Konzepte für eine ländliche Entwicklung*, (ed. V. Hoffmann), pp. 271–6. Josef Margraf, Weikersheim.
6 This section draws heavily on van den Ban, A.W. (1988) Whose messages? *Reading Rural Development Communications Bulletin*, 23.
7 Knowles, M.S. (1970) *The Modern Practice of Adult Education: Andragogy Versus Pedagogy.* Association Press, New York.
8 Röling, N.G. and Engel, P. (eds) (1991) I.T. from a knowledge system perspective: concepts and issues. In *The Edited Proceedings of the European Seminar on Knowledge Management and Information Technology* (D. Kuiper and N.G. Röling), p. 10. Department of Extension Science, Agricultural University, Wageningen.
9 Banta, G.R. and Jayasuriya, S.K. (1991) Economic analysis of new technologies. In: *Basic Procedures for Agro-Economic Research* (anonymous), revised edition, pp. 133–42. International Rice Research Institute, Los Baños.
10 Pretty, J., Guijt, I., Thompson, J. and Scoones, J. (1995) *A Trainers Guide for Participatory Learning and Action.* International Institute for Environment and Development, London.
Grndstaff, T.B. and Messerschmidt, D.A. (1995) *A Manager's Guide to the Use of Rapid Rural Appraisal.* FAO, Bangkok.

Mettrick, H. (1993) *Development Oriented Research in Agriculture.* International Centre for Development Oriented Research in Agriculture, Wageningen.
11 Warren, D.M. (1991) *Using indigenous knowledge in agricultural development.* World Bank Discussion Papers, 127, Washington DC.
12 Engel, P.G.H. (1995) *Facilitating Innovation: An Action-oriented Approach and Participatory Methodology to Improve Innovative Social Practice in Agriculture.* Agricultural University, Wageningen.
13 Axinn, G.H. (1988) *Guide on Alternative Extension Approaches.* FAO, Rome.
14 McDermott, J.K. (1987) Making extension effective: The role of research/extension linkages. In: *Agricultural Extension Worldwide; Issues, Practices and Emerging Priorities* (W.M. Rivera and S.G. Schram), pp. 89–99. Croom Helm, London.
15 Havelock, R.G. (1969) *Planning for Innovation Through the Dissemination and Utilization of Knowledge.* University of Michigan, Center for Research on the Utilization of Scientific Knowledge, Ann Arbor.
16 Green, L.W. and Kreuter, M.W. (1991) *Health Promotion Planning: An Educational and Environmental Approach.* Mayfield, Mountain View, California.
17 Rogers, E.M. (1973) *Communication Strategies for Family Planning.* Free Press, New York.
18 See Havelock, *op. cit.*
19 Schein, E.H. (1988) *Process Consultation: Its Role in Organization Development.* Addison Wesley, Reading.

Chapter 3

1 See the *Small Farmers' Development Manual* (1978), FAO, Bangkok.

Chapter 4

1 Proost, M.D.C. (1993) De dagen van de landbouwvoorlichter zijn geteld; Over veranderingen in het beroep van landbouwvoorlichter (The days of the agricultural extension agents have been counted; Changes in the extension profession). In: *Jaarboek Public Relations en Voorlichting 1993.* NGPR, The Hague.
2 Northouse, P.G. and Northouse, L.L. (1992) *Health Communication: Strategies for Health Professionals*, 2nd edn. Appleton and Lange, Norwalk.

Chapter 5

1 See discussion of visual literacy in Fuglesang, A. (1982) *About Understanding.* Hammarskjöld Foundation, Uppsala.
2 Dudley, E. and Haaland, A. (1993) *Communicating for Building Safety.* Intermediate Technology Publications, London.
3 We are grateful to Dr Niels Röling for his contribution to this section.
4 Berlo, D.K. (1960) *The Process of Communication.* Holt, Rinehart and Winston, New York.

5 Quoted in Frijda, N.J. (1965) *Kunnen mensen denken?* (Inaugural address.) University of Amsterdam, Amsterdam.
6 Flavier, J.M. (1993) *Doctor in the Barrio: Experiences with the Philippine Rural Reconstruction Movement.* New Day Publishers, Quezon City, p.167.
7 Hawkins, H.S. (1974) Visual Communication of Science and Technology. Chapter 6 in *Visual Education* (ed. C.E. Moorehouse). Pitman, Melbourne.
8 Ascroft, J. *et al.* (1973) *Extension and the forgotten farmer. Report of a field experiment.* Bulletin 37, Afdelingen voor Sociale Wetenschappen, Landbouwhogeschool, Wageningen.
9 Röling, N.G. and van der Fliert, E. (1994) The transformation of extension for sustainable agriculture: The case of Integrated Pest Management in Indonesia. *Agriculture and Human Values*, **11**: 96–108.
10 Ajzen, I. and Fishbein, M. (1980) *Understanding Attitudes and Predicting Social Behavior.* Prentice Hall, Englewood Cliffs. Also Ajzen, I. (1988) *Attitudes, Personality and Behavior.* Dorsey Press, Chicago.
11 See (1) McGuire, J. (1985) Attitudes and attitude change. In: *The Handbook of Social Psychology* (eds G. Lindzey and E. Aronson), Vol. 2. Random House, New York. (2) Kok, G. (1991/92) Health education theories and research for AIDS prevention. *Hygie*, **10**, 32–9.
12 Bos, A.H. (1974) *Oordeelsvorming in groepen.* Veenman, Wageningen.
13 Rogers, E.M. (1995) *Diffusion of Innovations*, 4th edn., p. 163. Free Press, New York.
14 Dasgupta, S. (1989) *Diffusion of Agricultural Innovations in Village India.* Wiley Eastern, New Delhi.
15 Fliegel, F.C. (1993) *Diffusion Research in Rural Sociology: The Records and Prospects for the Future*, pp. 72–87. Greenwood Press, Westport, Connecticut.
16 Rogers, E.M. (1976) *Communication and Development: Critical Perspectives.* Sage, Beverly Hills. Also published as a special edition of *Communication Research*, **3**(2).
17 See Zaltman, G. (1965) *Marketing: Contributions From the Behavioral Sciences*, pp. 107–116. Harcourt Brace and World, New York.
18 Jones, G.R. & Rolls, M. (eds) (1982) *Extension and Relative Advantage in Rural Development: Progress in Rural Extension and Community Development.* Wiley, Chichester.

Chapter 6

1 Many extension methods, including some we do not mention, are described in Sanders, H.C. *et al.* (eds) (1967) *The Cooperative Extension Service*, pp. 111–250. Prentice Hall, Englewood Cliffs.

Individual, group and mass media extension methods of relevance in less industrialized countries also are discussed in Swanson, B.E. (ed.) (1984) *Agricultural Extension: A Reference Manual*, pp. 130–55. FAO, Rome.
2 Klapper, J.T. (1960) *The Effects of Mass Communication.* Free Press, New York.
3 Windahl, S., Signitzer, B. and Olsen, J.T. (1991) *Using Communication Theory: An Introduction to Planned Change.* Sage, London.

4 Hoffmann, V. (1982) Intercultural Communication: The 'Cow-Case' and its Use in Training and Teaching. *Proceedings of the Fifth European Seminar on Extension Education*, University of Hohenheim, Stuttgart-Hohenheim.
5 We write 'can' here because it is not yet quite clear when media have this effect and when they do not. See: Weaver, D. (1981) Media Agenda-Setting and Media Manipulation. *Massacommunicatie*, **5**, 213–29.
6 Nair, K.S. and White, S.A. (1993) The development communication process: A reconceptualization. In: *Perspectives on Development Communication* (eds K.S. Nair and S.A. White). Sage, New Delhi.
7 See Schramm, W. and Lerner, D. (1976) *Communication and Change: The Last Ten Years – and the Next*. East–West Center, University of Hawaii, Honolulu.
8 This research is reviewed in Langer, I., Schulz von Thun, F. and Tausch, R. (1981) *Sich verständlich ausdrücken*. Reinhardt, München.
9 The use of pre-testing to improve communication is discussed in Dudley, E. and Haaland, A. (1991) *Communicating Building Safety*. Intermediate Technology Publications, London.
10 Bligh, D.E. (1972) *What's the Use of Lectures?* Penguin, Harmondsworth.
11 Warren, D.M. (1991) Using Indigenous Knowledge in Agricultural Development. World Bank Discussion Papers 127, Washington DC.
12 Lewin, K. (1953) Studies in Group Decision. In: *Group Dynamics, Research and Theory* (D. Cartwright and A. Zander). Row, Peterson Co., Evanston.
13 Campbell, C.A. (1995) Landcare: Participative Australian approaches to inquiry and learning for sustainability. *Journal of Soil & Water Conservation*, **50**(2), 125–31.
14 For a comparison of directive and non-directive methods to lead group discussions see Batten, T.R. (1967) *The Non-Directive Approach in Group and Community Work*. Oxford University Press, London.
15 Brammer, L.M. (1973) *The Helping Relationship*, pp. 21–7. Prentice Hall, Englewood Cliffs.
16 Rogers, C.R. (1962) The interpersonal relationship: the core of guidance. *Harvard Educational Review*, **32**, 416–529.
17 See research reports by: Cook, B.L. (1981) *Understanding Pictures in Papua New Guinea*. David Cook Foundation, Elgin, Illinois; Dudley, E. and Haaland, A. (1991) *Communicating Building Safety*. Intermediate Technology Publications, London.
18 Peter, I. (1995) Use of computer information systems by Landcare groups for technical support, Chapter 16 in: *Participative Approaches for Landcare: Perspectives, Policies, Programmes* (eds S. Chamala and K. Keith). Australian Academic Press, Department of Primary Industry, Brisbane.

Chapter 7

1 See Anonymous (1988) *ZOPP (An Introduction to the Method)*. Deutsche Gesellschaft für Technische Zusammenarbeit (GTZ), Eschborn.
2 Moris, J. and Copestake, J. (1993) *Qualitative Enquiry for Rural Development*. Intermediate Technology Publications, London.
3 See Wilkening, E.A. (1958) Consensus in role definition of county extension agents between agents and local sponsoring committee members. *Rural Sociology*, **23**,

184–97; and Beal, G.M. and Blount, R.C. *et al.* (1966) *Social Action and Interaction in Program Planning*. Iowa State University Press, Ames.

Chapter 8

1 This is based substantially on Bennett, C.F. (1976) *Analysing Impacts of Extension Programs*. USDA Extension Service, Washington DC.
2 See Dunn, W.N. (1982) Reforms as arguments. *Knowledge*, **3**, 293–326. (In a stimulating article Dunn discusses several other limitations of the use of experiments.) Schön, D.A., Drake, W.D. and Miller, R.A. (1984) Social experimentation as reflection-in-action. *Knowledge*, **6**, 5–35 (present an alternative).
3 See Weiss, C.H. (1975) Interviewing in evaluation research. In: *Handbook of Evaluation Research* (E.L. Struening and M. Guttentag), pp. 355–95. Sage, Beverly Hills.

Chapter 9

1 See Oakley, P. (1991) *Projects with People: The Practice of Participation in Rural Development*. International Labour Organization (ILO), Geneva.
2 Hussi, P., Murphy, J., Lindberg, O. and Brenneman, L. (1993) *The development of cooperatives and other rural organizations*. World Bank Technical Paper, 199, xv, Washington DC.
3 Burkey, S. (1993) *People First: A Guide to Self-reliant Participatory Rural Development*, p. 165. Zed Books, London.
4 Hayward, J.A. (1989) World Bank involvement in agricultural extension. In: *Technology Systems for Small Farmers* (ed. A.M. Kesseba), pp. 133–52. Westview, Boulder.
5 For example, see *Forests, Trees and People Newsletter*, 22, November 1993, Uppsala.

Chapter 10

1 In the recent extension literature there is much discussion, often without agreement, on the changes which are needed. For example, see

Antholt, C.H. (1994) *Getting ready for the twenty-first century; technical change and institutional modernization in agriculture*. World Bank Technical Paper, 217, Washington DC.
Röling, N.G. and Jiggins, J.L.S. (1994) Extension and the sustainable management of natural resources. *European Journal of Agricultural Education and Extension*, **1**(1), 23–43.
Scoones, I. and Thompson, J. (eds) (1994) *Beyond Farmer First; Rural People's Knowledge, Agricultural Research and Extension Practice*. Intermediate Technology Publications, London.

2 We can only make broad generalizations here, some of which will not be true in the work situation of all readers. We hope that this will help the reader to analyse his own situation.

3 Leonard, D.K. (1977) *Reaching the Peasant Farmer: Organization Theory and Practice in Kenya*, Chapter 5, University of Chicago Press, Chicago.

4 Antholt, C.H. (1994) *Getting ready for the twenty-first century; technical change and institutional modernization in agriculture*. World Bank Technical Paper, 217, Washington DC: 28.

5 Martwanna, N. and Chamala, S. (1991) Training for rural development in Thailand: content or process model: *Community Development Journal*, **26**, 43–9.

6 Herzberg, F. (1968) *Work and the Nature of Man*. Staples Press, London.

7 Evans, J.F. and Dahl, D.T. (1984) Organizing for extension communications. In: *Agricultural Extension* (ed. B.E. Swanson). FAO, Rome.

8 Several publications give useful information for planning extension programmes for farm women. For example:

Saito, K.A. and Spurling, D. (1992) *Developing agricultural extension for women farmers*. World Bank Discussion Paper, 156, Washington DC.

9 The extension service of the Department of Agriculture in the Australian state of Tasmania commenced charging fees for certain advisory services in 1982. Initial reaction from farmers was to make less use of the service for situations in which fees are charged (as in personal visits) and more use of those services for which no fees are charged (as in group activities).

10 Benor, D. and Baxter, M. (1984) *Training and Visit Extension*. World Bank, Washington DC.

11 John, K.C., Rajan, C.S., Singh, C. and Arora, S.K. (1993) *Farmers' Participation in Agricultural Research and Extension Systems*. MANAGE (National Institute of Agricultural Extension Management), Hyderabad.

12 Antholt, C.H. (1994) *Getting ready for the twenty-first century; technical change and institutional modernization in agriculture*. World Bank Technical Paper, 217, Washington DC.

13 ISNAR (International Service for National Agricultural Research) has published many studies of the research–extension linkage. These are summarized in Merrill-Sands, D. and Kaimowitz, D. (undated) *The Technology Triangle; Linking Farmers, Technology Transfer Agents and Agricultural Researchers*. ISNAR, The Hague.

Chapter 11

1 See Bardsley, J.B. (1982) *Farmers' Assessment of Information and its Sources*. School of Agriculture and Forestry, University of Melbourne.

Index